A LEVEL
Questions and Answers

MECHANICS

Michael Jennings & Bronwen Moran

Principal Examiners

Letts
EDUCATIONAL

SERIES EDITOR: BOB McDUELL

Contents

Introduction

HOW TO USE THIS BOOK

The aim of the *Questions and Answers* series is to provide students with the help required to attain the highest level of achievement in important examinations. This book is intended to help you with the Mechanics component of A- and AS-level Mathematics. The series relies on the idea that an experienced examiner can provide, through examination questions, sample answers and advice, the help students need to secure success. Many revision aids concentrate on providing factual information that might have to be recalled in an examination. This series, while giving factual information in an easy-to-remember form, concentrates on the other skills that need to be developed for the new A-level examinations being introduced from 1996.

The *Questions and Answers* series is designed to provide:

- Easy-to-use **Revision Summaries** that identify important factual information that students must understand if progress is to be made in answering examination questions.

- Advice on the different types of question in each subject and how to answer them well to obtain the highest marks.

- Information about other skills, apart from the recall of knowledge, that will be tested on examination papers. These are sometimes called **assessment objectives** and modern A-level examinations put great emphasis on them. The *Questions and Answers* series is intended to develop these skills, particularly of communication, problem-solving, evaluation and interpretation, by the use of questions and the appreciation of outcomes by the student.

- Many examples of **examination questions**. Students can increase their achievement by studying a sufficiently wide range of questions, provided that they are shown the way to improve their answers to these questions. It is advisable that students try the questions first before looking at the answers and the advice that accompanies them. All the Mathematics questions come from actual examination papers or specimen materials issued by the British Examination Boards, reflecting their requirements.

- **Sample answers** and mark schemes to all the questions.

- **Advice from Examiners**: by using the experience of actual examiners we are able to give advice that can enable students to see how their answers can be improved to ensure greater success.

Success in A-level examinations comes from proper preparation and a positive attitude, developed through a sound knowledge of facts and an understanding of principles. These books are intended to overcome 'examination nerves' which often come from a fear of not feeling properly prepared.

THE IMPORTANCE OF USING QUESTIONS FOR REVISION

Past examination questions play an important part in revising for examinations. However, it is important not to start practising questions too early. Nothing can be more disheartening than trying to do a question that you do not understand because you have not mastered the concepts. Therefore it is important to have studied a topic thoroughly before attempting questions on it.

It is unlikely that any question you try will appear in exactly the same form on the papers you are going to take. However the number of totally original questions that can be set on any part of the syllabus is limited and so similar ideas occur over and over again. It certainly will help you if the question you are trying to answer in an examination is familiar and you are used to the type of language used. Your confidence will be boosted, and confidence is important for examination success.

Practising examination questions will also highlight gaps in your knowledge and understanding that you can go back and revise more thoroughly. It will indicate which sorts of question you can do well and which, if there is a choice, you should avoid.

Finally, having access to answers, as you do in this book, will enable you to see clearly what is required by the examiner, how best to answer each question and the amount of detail required. Remember that attention to detail is a very important aspect of achieving success at A-level.

MAXIMISING YOUR MARKS IN MATHEMATICS

One of the keys to examination success is to know how marks are gained or lost and the examiner's tips given with the solutions in this book give hints on how you can maximise your marks on particular questions. However you should also take careful note of these general points:

- Check the requirements of your examination board and follow the instructions (or 'rubric') carefully about the number of questions to be tackled. Many A-level Mathematics examinations instruct you to attempt all the questions and where papers start with short, straightforward questions, you are advised to work through them in order so that you build up your confidence. Do not overlook any parts of a question – double-check that you have seen everything, including any questions on the back page! If there is a choice, do the correct number. If you do more, you will not be given credit for any extra and it is likely that you will not have spent the correct time on each question and your answers could have suffered as a result. Take time to read through all the questions carefully, and then start with the question you think you can do best.

- Get into the habit of setting out your work neatly and logically. If you are untidy and disorganised you could penalise yourself by misreading your own figures or lose marks because your method is not obvious. Always show all necessary working so that you can obtain marks for a correct method even if your final answer is wrong. Remember that a good clear sketch can help you to see important details.

- When the question asks for a particular result to be established, remember that to obtain the method marks you must show sufficient working to convince the examiner that your argument is valid. Do not rely too heavily on your graphical calculator.

- Do not be sloppy with algebraic notation or manipulation, especially when using brackets and negatives. Do rough estimates of calculations to make sure that they are reasonable, state units if applicable and give answers to the required degree of accuracy; do not approximate prematurely in your working.

- Make sure that you interpret your answers in the context of the questions and be careful that your answers make sense in that context.

- Make sure that you are familiar with the formulas booklet and tables that you will be given in the examination and learn any useful formulas that are not included. Refer to the booklet in the examination and transfer details accurately.

- When about 15 minutes remain, check whether you are running short of time. If so, try to score as many marks as possible in the short time that remains, concentrating on the easier parts of any questions not yet tackled.

- The following glossary may help you in answering questions:
  Write down, state – no justification is needed for an answer.
  Calculate, find, determine, show, solve – include enough working to make your method clear.
  Deduce, hence – make use of the given statement to establish the required result.
  Sketch – show the general shape of a graph, its relationship with the axes, any asymptotes and points of special significance such as turning points.
  Draw – plot accurately, using graph paper and selecting a suitable scale; this is usually preparation for reading information from the graph.
  Find the <u>exact</u> value – leave it in fractions or surds, or in terms of logarithms, exponentials or π; note that using a calculator is likely to introduce decimal approximations, resulting in loss of marks.

The quantities which occur in mechanics fall into two classes. Firstly, there are those which can be completely specified by their magnitude, such as mass, length and time; these are called **scalars**. Secondly, there are quantities, such as force, displacement and velocity, which require a statement of their direction as well as of their magnitude for their complete specification; these are called **vectors**.

If we take any two points A and B, then the length AB and the direction from A to B together constitute a vector which can be denoted by $\overrightarrow{AB}$ or by a single symbol **a**, written in bold type.

Vector addition

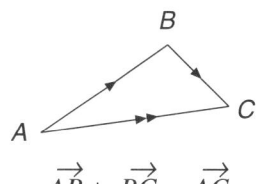

$$\overrightarrow{AB} + \overrightarrow{BC} = \overrightarrow{AC}$$

Scalar multiplication

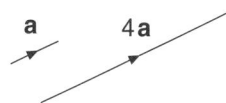

Vector subtraction

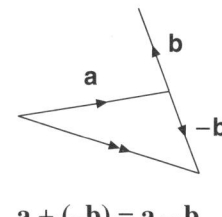

$$\mathbf{a} + (-\mathbf{b}) = \mathbf{a} - \mathbf{b}$$

Magnitude or length of a vector

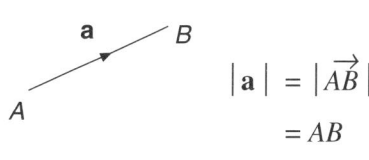

$$|\mathbf{a}| = |\overrightarrow{AB}|$$
$$= AB$$

a is a unit vector if it has length 1, i.e., $|\mathbf{a}| = 1$.

i, **j** and **k** are the unit vectors along the x-axis, the y-axis and the z-axis, and are sometimes called **base vectors**.

If O is the origin and P is any point, then $\overrightarrow{OP}$ is called the **position vector** of P relative to O.

If P has coordinates (x, y, z), then its position vector, $\overrightarrow{OP}$, is given by

$$\overrightarrow{OP} = x\mathbf{i} + y\mathbf{j} + z\mathbf{k} \text{ and } |\overrightarrow{OP}| = OP = \sqrt{x^2 + y^2 + z^2}.$$

Note that some Exam Boards write $x\mathbf{i} + y\mathbf{j} + z\mathbf{k}$ in column vector form $\begin{pmatrix} x \\ y \\ z \end{pmatrix}$.

If P has position vector **p** and Q has position vector **q**, then $\overrightarrow{PQ} = \mathbf{q} - \mathbf{p}$

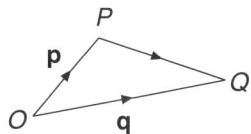

$$\overrightarrow{PQ} = \overrightarrow{PO} + \overrightarrow{OQ}$$
$$= -\mathbf{p} + \mathbf{q}$$
$$= \mathbf{q} - \mathbf{p}$$

The **unit vector** in the direction of **p**, $\hat{\mathbf{p}}$, is given by $\hat{\mathbf{p}} = \dfrac{\mathbf{p}}{|\mathbf{p}|}$, so if $\mathbf{p} = x\mathbf{i} + y\mathbf{j} + z\mathbf{k}$ then

$$\hat{\mathbf{p}} = \left(\frac{\mathbf{p}}{\sqrt{x^2 + y^2 + z^2}} \right)$$

If **p** is any vector, then the vector $k\hat{\mathbf{p}}$ is a vector of magnitude k along the direction of **p**.

REVISION SUMMARY

Resolving a vector into components

Any vector **F** may be **resolved** into (usually two perpendicular) **components**. The component of **F** in a particular direction D gives the magnitude of the effect of **F** in that direction and is given by $F\cos\theta$ where F is the magnitude of **F** and θ is the angle between the direction of **F** and the direction D. Note that the component perpendicular to the direction D is given by $F\cos(90° - \theta)$ which is $F\sin\theta$.

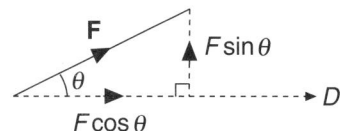

Scalar product

The definition of the scalar product is

$$\mathbf{a.b} = ab\cos\theta$$

where $a = |\mathbf{a}|$ and $b = |\mathbf{b}|$ and θ is the angle between **a** and **b**.

The scalar (or dot) product **a.b** is so-called because its value is a scalar, i.e., a number, and the operation behaves like an ordinary product, i.e. it obeys the distributive law:

$$\mathbf{a.(b + c) = a.b + a.c}$$

One of the most important properties of the scalar product is the following:

If **a** and **b** are non-zero vectors then **a.b** = 0 is equivalent to **a** and **b** being perpendicular, since $\cos 90° = 0$. Also $\mathbf{a.a} = a^2$, where $a = |\mathbf{a}|$, since $\cos 0° = 1$.

Using these results we obtain $\mathbf{i.i = j.j = k.k} = 1$ and $\mathbf{i.j = j.k = i.k} = 0$ and so if

$\mathbf{a} = a_1\mathbf{i} + a_2\mathbf{j} + a_3\mathbf{k}$ and $\mathbf{b} = b_1\mathbf{i} + b_2\mathbf{j} + b_3\mathbf{k}$, then using the distributive law,

$$\mathbf{a.b} = (a_1\mathbf{i} + a_2\mathbf{j} + a_3\mathbf{k}).(b_1\mathbf{i} + b_2\mathbf{j} + b_3\mathbf{k}) = a_1b_1 + a_2b_2 + a_3b_3$$

To find the angle θ between the vectors **a** and **b** we can use

$$\cos\theta = \frac{\mathbf{a.b}}{ab} = \frac{a_1b_1 + a_2b_2 + a_3b_3}{\sqrt{(a_1{}^2 + a_2{}^2 + a_3{}^2)}\sqrt{(b_1{}^2 + b_2{}^2 + b_3{}^2)}} \qquad [1]$$

Vector equation of a straight line

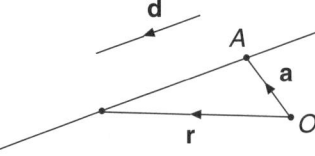

If you need to revise this subject more thoroughly, see the relevant topics in the *Letts* A-level *Mathematics Study Guide*.

A straight line passes through a given point A, with position vector **a**, and is parallel to the vector **d**. The position vector, **r**, of a general point on the line is given by $\mathbf{r = a} + t\mathbf{d}$, where t is a parameter. This is known as the **vector equation of the line**. In Mechanics, if **a** was the position vector of the initial position of a ship moving with constant velocity **d**, then **r** would be its position vector at time t, and the straight line would be the path taken by the ship.

To find the angle between two straight lines given in vector form, find the angle between their direction vectors, using [1] above.

To find the point of intersection of two lines using parameters s and t in their equations, set up three simultaneous equations in s and t by equating coefficients of **i**, **j** and **k**, and solve for s and t, checking carefully that your values work in all three equations – if so, then the two lines do intersect.

QUESTIONS

1 Two forces, of magnitudes 5 N and 7 N, act on a body. The angle between the forces is $55°$. The resultant has magnitude R and acts at an angle θ to the 5 N force, as shown in the diagram. Calculate R and θ.

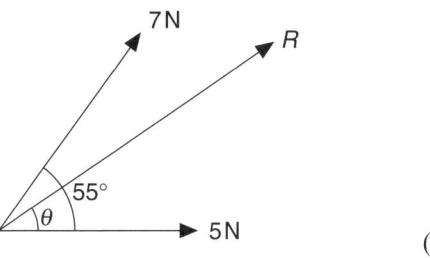

(6)

UCLES

2 A boy is attempting to take two dogs for a walk. The dogs exert horizontal forces of 40 N and 50 N on the boy in directions making $120°$ with each other. By modelling the boy as a particle find the magnitude and direction of the horizontal force which the boy must exert to maintain equilibrium. (11)

NICCEA

3 (a) Given that $\mathbf{a} = 2\mathbf{i} + \mathbf{j} + q\mathbf{k}$ and $\mathbf{b} = q\mathbf{i} - 2\mathbf{j} + 2q\mathbf{k}$

   (i) Find the values of q such that $\mathbf{a}$ and $\mathbf{b}$ have equal magnitudes. (4)

   (ii) Given that $q = 3$, find the resultant of $\mathbf{a}$ and $\mathbf{b}$. (2)

   (iii) Given that $q = 2$, find a unit vector in the direction of $\mathbf{a}$. (3)

   (b) Three forces, each of magnitude F, act along the sides of an equilateral triangle as shown in the figure. Find the magnitude and direction of their resultant.

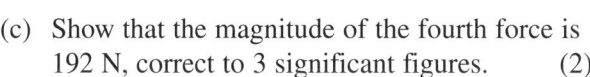

(4)

NICCEA

4 Three cables exert forces that act in a horizontal plane on the top of a telegraph pole.

   (a) Find the resultant of these 3 forces, in terms of the unit vectors $\mathbf{i}$ and $\mathbf{j}$. (2)

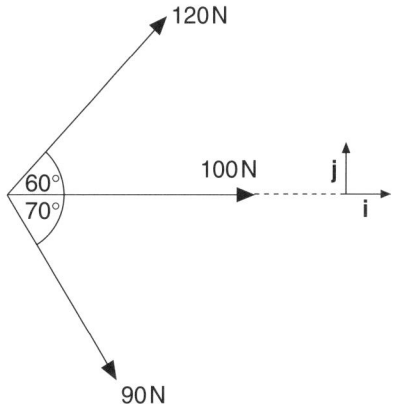

   (b) A fourth cable is attached to the top of the telegraph pole to keep the pole in equilibrium. Find the force, exerted by this fourth cable, in terms of $\mathbf{i}$ and $\mathbf{j}$. (1)

   (c) Show that the magnitude of the fourth force is 192 N, correct to 3 significant figures. (2)

   (d) On a diagram show clearly the direction in which the fourth force acts. (2)

   (e) The fourth cable does not lie in the same horizontal plane as the other 3 cables and the tension in this cable is in fact 200 N. Find the angle between this cable and the horizontal plane. (2)

AEB

1 Vectors

5 (a) Determine whether the point (1, 3, 7) lies on the straight line between (–1, 4, 5) and (5, 2, 11).   (2)

(b) Calculate in radians the angle between the vectors $\begin{pmatrix} 2 \\ 3 \\ 4 \end{pmatrix}$ and $\begin{pmatrix} -3 \\ 0 \\ 1 \end{pmatrix}$.   (2)

(c) Write down an equation linking the vectors **p**, **q** and **r** in the figure.

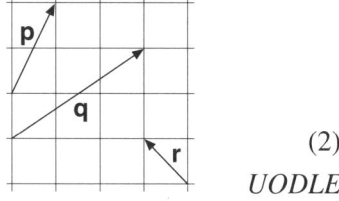

                                                                                                                            (2)

UODLE

6 (a)

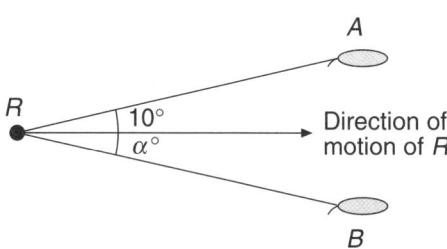

Two cart-horses, *A* and *B*, pull a small heavy rock *R* in a straight line over rough horizontal ground by means of two ropes attached to the rock. The horses are separated from each other so that each rope makes a small angle with the direction of motion of the rock, as shown in the figure.

Throughout the motion, the rope attached to *A* has a tension of 800 N and makes an angle of 10° with the direction of motion of *R*, and the rope attached to *B* has a tension of 500 N and makes an angle of $\alpha°$ with the direction of motion of *R*. The rock is modelled as a particle, the ropes are assumed to be horizontal, and the air resistance is assumed to be negligible. Using this model,

(a) find, to the nearest whole number, the value of α.

Given that the horses drag the rock very slowly along the ground at a constant speed,

(b) find, to 3 significant figures, the resistance to motion experienced by the rock.

(c) Suggest one reason why it is reasonable to ignore air resistance in the situation described.

(d) Suggest one refinement of the model, in relation to the ropes, which should be incorporated to make the model a more accurate reflection of the situation.   (9)

London Examinations

7 Two forces $\mathbf{F}_1 = (2\mathbf{i} + 3\mathbf{j})$ N and $\mathbf{F}_2 = (\lambda\mathbf{i} + \mu\mathbf{j})$ N, where λ and μ are scalars, act on a particle. The resultant of the two forces is **R**, where **R** is parallel to the vector $\mathbf{i} + 2\mathbf{j}$.

(a) Find, to the nearest degree, the acute angle between the line of action of **R** and the vector **i**.

(b) Show that $2\lambda - \mu + 1 = 0$.

Given that the direction of $\mathbf{F}_2$ is parallel to **j**,

(c) find, to 3 significant figures, the magnitude of **R**.   (9)

London Examinations

4

8 A football pitch is a horizontal plane and *O* is a fixed point on the pitch. The vectors **i** and **j** are perpendicular unit vectors in this horizontal plane. Colin and David are two players on the pitch. At time $t = 0$, David kicks the ball from the origin *O* with a constant velocity $8\mathbf{i}$ m s^{-1} and runs thereafter with constant velocity $(3\mathbf{i} + 5\mathbf{j})$ m s^{-1}. When David kicks the ball, Colin is at the point with position vector $(10\mathbf{i} + 8\mathbf{j})$ m and starts running with constant velocity $(3\mathbf{i} - 4\mathbf{j})$ m s^{-1}.

(a) Write down the position vectors of Colin and David at time *t* seconds.

(b) Verify that Colin intercepts the ball after 2 seconds.

As soon as Colin intercepts the ball he kicks it, giving it a constant velocity of $(\lambda\mathbf{i} - \mu\mathbf{j})$ m s^{-1}. He aims to pass it to David who maintains his constant velocity. Given that David intercepts the ball 2 seconds after Colin has kicked it,

(c) find the values of λ and μ. (14)

London Examinations

9

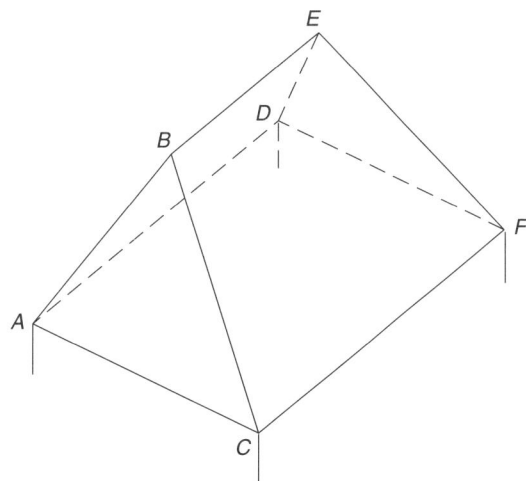

The diagram shows an architect's drawings of the roof of a house. The lines *AD*, *BE* and *CF* are parallel. In the architect's model, the lines *BE* and *BC* have vector equations

$$\mathbf{r} = \begin{pmatrix} 3 \\ 4 \\ 7 \end{pmatrix} + \lambda \begin{pmatrix} 1 \\ 0 \\ 0 \end{pmatrix}$$

and

$$\mathbf{r} = \begin{pmatrix} 10 \\ 0 \\ 5 \end{pmatrix} + \mu \begin{pmatrix} -3 \\ 4 \\ 2 \end{pmatrix},$$

respectively.

(a) Find the coordinates of the point *B*. (2)

(b) Given that the coordinates of the point *C* are (10, 0, 5), write down a vector equation of the line *CF*. (1)

(c) Given further that the coordinates of the point *A* are (10, 8, 5),

  (i) find a vector equation of the line *CA*, (2)

  (ii) by using an appropriate scalar product, or otherwise, find the angle *BCA*, giving your answer to the nearest tenth of a degree. (3)

NEAB

2 *Kinematics of a particle*

The study of the motion of a particle without considering the forces that are acting upon it is known as **Kinematics**.

If the particle is moving with **constant acceleration** (it may be given as a numerical constant in a given direction or in vector form, e.g. $(2\mathbf{i} - \mathbf{j})\,\mathrm{m\,s^{-2}}$) then we can use one or more of the following formulae to analyse its motion:

$$\mathbf{v} = \mathbf{u} + \mathbf{a}t \qquad \mathbf{s} = \mathbf{u}t + \tfrac{1}{2}\mathbf{a}t^2 \qquad \mathbf{s} = \tfrac{1}{2}(\mathbf{u} + \mathbf{v})t \qquad \mathbf{v}^2 = \mathbf{u}^2 + 2\mathbf{a}\mathbf{s} \qquad \mathbf{s} = \mathbf{v}t - \tfrac{1}{2}\mathbf{a}t^2$$

Notice that there are five quantities: $\mathbf{a}$, $\mathbf{u}$, $\mathbf{v}$, $\mathbf{s}$ and t and each equation gives a relationship between four of them. Notice also that the first four are all vectors. If you are given values of these in numerical form then before you apply one of the formulae you will need to choose a positive direction – if you are given values in $\mathbf{i}$-$\mathbf{j}$ form then they can be substituted into a formula in that form. Before starting a problem it is a good idea to write down which of the five quantities you know and which you don't – this will help you to decide which of the formulae to apply.

If the particle is moving with **variable acceleration** you will need to use **calculus** to analyse its motion; displacement, velocity and acceleration are all related:

$$\mathbf{s} \qquad\qquad \mathbf{v} = \frac{d\mathbf{s}}{dt} \qquad\qquad \mathbf{a} = \frac{d\mathbf{v}}{dt}$$

$$\mathbf{s} = \int \mathbf{v}\,dt \qquad \mathbf{v} = \int \mathbf{a}\,dt \qquad \mathbf{a}$$

$$\text{-----} \blacktriangleright \quad differentiate \quad \text{-----} \blacktriangleright$$

$$\text{displacement} \qquad \text{velocity} \qquad \text{acceleration}$$

$$\blacktriangleleft \text{------} \quad integrate \quad \blacktriangleleft \text{------}$$

Note: $\mathbf{s}$ is often given as the **position vector r** (i.e. displacement relative to the origin O).

Also $a = \dfrac{dv}{dt} = \dfrac{ds}{dt} \times \dfrac{dv}{ds} = v\dfrac{dv}{ds}$ by the Chain Rule.

Displacement–time, velocity–time and acceleration–time graphs

The gradient of a displacement–time graph gives the average velocity over a given time, if it is calculated between two points on the graph, or the velocity at a particular instant if it is calculated at a particular point – if the graph is a curve this will mean finding the gradient of the tangent to the curve at that particular point. Note that this could be calculated by finding the value of ds/dt at the point concerned.

Similarly, the gradient of a velocity–time graph gives the average acceleration over a given time if it is calculated between two points on the graph, or the acceleration at a particular instant if it is calculated at a particular point. Note that this could be calculated by finding the value of dv/dt at the point concerned. Also the area under a velocity–time graph gives the distance travelled. Velocity–time graphs are particularly useful for solving problems in which the motion consists of several stages, e.g. if a car accelerates then moves with constant velocity and then decelerates.

For **projectile** motion the usual approach is **to consider the horizontal and vertical motion separately and then use the constant acceleration formulae given above**. This is justified because when a body of mass m is moving through the air, our modelling assumption at this level is that the only force acting on it is its **weight mg** vertically downwards (i.e. we ignore the effects of air resistance, spin and variation in g). Thus its **acceleration is g, a constant, taken to be 9.8, 9.81 or $10\,\mathrm{m\,s^{-2}}$, vertically downwards.** In order to consider the horizontal and vertical motion separately we must consider each of the quantities $\mathbf{a}$, $\mathbf{u}$, $\mathbf{v}$ and $\mathbf{s}$ and their horizontal and vertical components.

For the motion from O to A, for example:

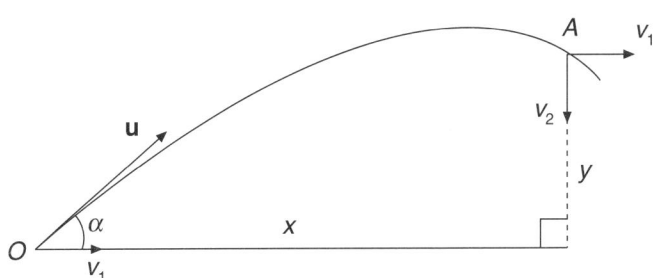

	Horizontal component	Vertical component
a	0	$-g$
u	$u\cos\alpha$	$u\sin\alpha$
v	v_1	$-v_2$
s	x	y

REVISION SUMMARY

We can then apply any of the formulae either horizontally or vertically, e.g.

$\mathbf{s} = \mathbf{u}t + \frac{1}{2}\mathbf{a}t^2$ vertically: $\qquad y = u\sin\alpha t + \frac{1}{2}(-g)t^2 \qquad$ [1]

$\mathbf{s} = \mathbf{u}t + \frac{1}{2}\mathbf{a}t^2$ horizontally: $\qquad x = u\cos\alpha t \qquad$ [2]

If we eliminate t between these two equations we can obtain the **equation of the path**

or trajectory of the projectile: $\quad y = x\tan\alpha - \dfrac{gx^2\sec^2\alpha}{2u^2}$

This formula is particularly useful if we are given the initial projection speed u and a 'target' whose x and y values are given and we wish to find the possible angle(s) of projection which will ensure that the target is hit. The method is to write $\sec^2\alpha$ as $(1 + \tan^2\alpha)$, thus obtaining a quadratic equation in $\tan\alpha$ which will have 0, 1 or 2 roots depending on the position of the target in relation to the projection point.

Time of flight, range and greatest height

Consider a projectile fired with speed U at an angle α to the horizontal from a point O on a horizontal plane which lands on the plane at the point A.

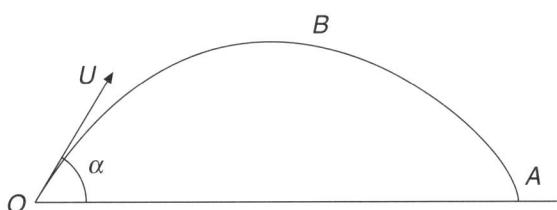

If we put $y = 0$ in equation [1] and solve for t, we obtain the time of flight $T = \dfrac{2U}{g}\sin\alpha$.

If we then use this value for t in equation [2], we obtain the range $R = \dfrac{2U^2}{g}\sin\alpha\cos\alpha$.

Writing $R = \dfrac{U^2}{g}\sin 2\alpha$, then $\boldsymbol{R_{max}} = \dfrac{U^2}{g}$ **when** $\sin 2\alpha = 1$, **i.e. when** $\alpha = 45°$.

Notice also that if we replace α by $(90° - \alpha)$ in the range formula we obtain the same result for the range, i.e. **projection at α to the horizontal gives the same horizontal range as projection at α to the vertical**.

Finally if we put $t = \dfrac{U}{g}\sin\alpha$ (half the time of flight) in equation [1] and simplify, we obtain the

greatest height, $H = \dfrac{U^2}{2g}\sin^2\alpha$. This result can be obtained more easily as follows:

Using $v^2 = u^2 + 2as$ vertically from O to B, the highest point on the path, $0^2 = U^2\sin^2\alpha + 2(-g)H$

which gives: $\quad H = \dfrac{U^2}{2g}\sin^2\alpha$

If you need to revise this subject more thoroughly, see the relevant topics in the *Letts* A-level Mathematics Study Guide.

1 A particle P of mass 0.1 kg moves so that, at time t seconds, its acceleration $\mathbf{a}\,\text{m s}^{-2}$ is given by

$$\mathbf{a} = 5t\mathbf{i} - 15t^{\frac{1}{2}}\mathbf{j}, \; (t \geq 0),$$

where the unit vectors $\mathbf{i}$ and $\mathbf{j}$ are directed due East and North respectively.

(a) Find the magnitude of the resultant force acting on P when $t = 4$, giving your answer in N to 2 decimal places.

Given that P has velocity $10\mathbf{i}\,\text{m s}^{-1}$ when $t = 0$, find

(b) an expression for the velocity of P in terms of t,

(c) the direction of motion of P when $t = 4$, giving your answer as a bearing to the nearest degree. (14)

London Examinations

2

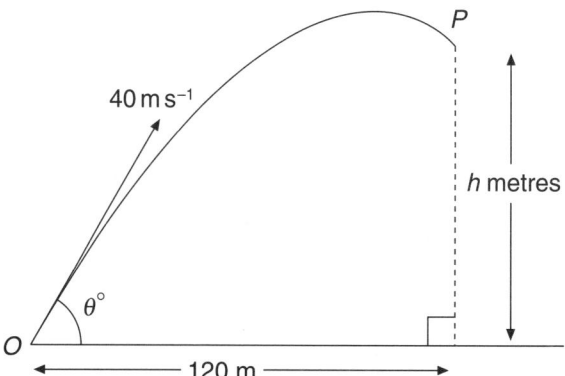

A golfer strikes a ball in such a way that it leaves the point O at an angle of elevation θ° and with a speed of $40\,\text{m s}^{-1}$. After 5 seconds the ball is at point P which is 120 m horizontally from O and h metres above O, as shown in the diagram. Assuming that the flight of the ball can be modelled by the motion of a particle with constant acceleration, find the value of θ and the value of h. (6)

Give one force on the golf ball that this model does not allow for. (1)

UCLES

3 A ball is bouncing on the hard surface of a horizontal playground. Initially the ball is dropped from rest at a height of 1.8 metres. Assume that while the ball is in the air the only force acting on it is that of gravity, and take the value of g to be $10\,\text{m s}^{-2}$.

(a) Calculate

(i) the time taken for the ball to fall from its initial position to the ground, (2)

(ii) the speed with which the ball first reaches the ground. (2)

(b) Each time the ball hits the ground it rebounds instantaneously with a speed which is half the speed with which it hit the ground. Calculate

(i) the time between its first and second impacts with the ground, (2)

(ii) the greatest height reached by the ball between its first and second impacts with the ground. (2)

(c) Name one naturally occurring force which might make the motion of the ball not the same as that predicted by your calculations. (1)

(continued)

(d) Imagine that the motion (with the same ball, playground and initial height) took place on the Moon, where the acceleration due to gravity is about a sixth of the value given above. For each of the quantities found in parts (a) (i), (a) (ii), (b) (i) and (b) (ii), state, with reasons, whether the value on the Moon would be the same as, greater than or less than the value on the Earth. (5)

NEAB

4 In this question distances are measured in metres and positions are expressed relative to an origin O. The unit vectors **i** and **j** are in the directions East and North respectively.

A radio-controlled model boat was put in a pond at the point O with initial velocity $0.6\mathbf{j}\,\mathrm{m\,s^{-1}}$ and its velocity after 15 seconds was measured as $(10.5\mathbf{i} - 0.9\mathbf{j})\,\mathrm{m\,s^{-1}}$. The acceleration of the boat is modelled as being constant.

(i) Show that the acceleration of the boat was $(0.7\mathbf{i} - 0.1\mathbf{j})\,\mathrm{m\,s^{-2}}$. (3)

(ii) Find an expression for the position of the boat at time t. For what value of t was the boat North-East of O? (4)

(iii) For what value of t was the boat travelling in the direction North-East? (4)

(iv) The boat lost contact with the radio transmitter when it was 100 m from O. Assuming that the acceleration remained constant, **verify** that contact was lost before 17 seconds had elapsed. (3)

Oxford & Cambridge

5 A railway timetabler has to determine the time taken between two stations a distance of 22.5 km apart. In order to do this he assumes that a train, on leaving one station, accelerates at a constant rate for 75 s until it reaches a speed of $30\,\mathrm{m\,s^{-1}}$ at a point A. It then continues at this speed to a point B when a constant retardation is applied for 75 s so that the train comes to rest in the second station.

(a) Draw a sketch of the velocity–time graph showing the motion of the train. (1)

(b) Find the magnitude of the acceleration and the distance travelled whilst the train is accelerating. (3)

(c) Find the total time estimated by the timetabler for the journey. (2)

(d) The timetabler realises that in making a timetable he has to insert a safety margin by allowing for the train having to stop at a signal. He does this by assuming that some time after passing A the train is forced to slow down to rest and it stops for 60 s. He assumes that the retardation and acceleration applied are the same as those applied at the start and end of the journey. He also assumes that the point where the train starts slowing down is such that the train will have resumed the constant speed of $30\,\mathrm{m\,s^{-1}}$ before reaching B. Find the total time that would now be estimated for the journey. (3)

WJEC

6 A dot moves on the screen of an oscilloscope so that its position relative to a fixed origin is given by:

$$r = 2t\,\mathbf{i} + \sin\left(\frac{\pi t}{2}\right)\mathbf{j}$$

(a) Sketch the path of the dot for $0 \le t \le 4$. (2)

(b) Find the velocity and acceleration of the dot when $t = 3$. Draw vectors on your diagram to show these two quantities. (6)

AEB

7

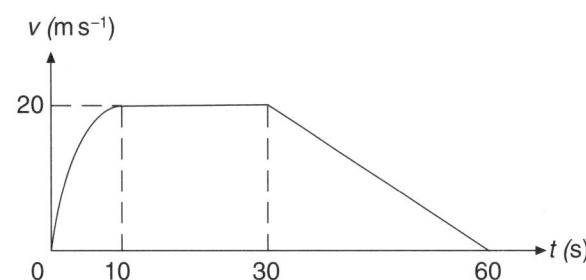

The (t, v) graph for a motor-cyclist travelling on a straight course is shown in the diagram, for $0 \le t \le 60$, where t is the time measured in seconds and v is the velocity measured in $m\,s^{-1}$. For $10 \le t \le 60$ the graph consists of two straight line segments. Calculate the distance travelled by the motor-cyclist during the period that he was slowing down. (2)

Use the (t, v) graph to show that the total distance travelled by the motor-cyclist during the 60 seconds is greater than 800 m. (2)

Explain how you could use the (t, v) graph to estimate the time at which the motor-cyclist was accelerating at $2\,m\,s^{-2}$. (2)

UCLES

8   A football is kicked through the air above a horizontal pitch. Unit vectors **i** and **j** are aligned horizontally and vertically upwards respectively. The ball's position from the point at which it was kicked, before it bounces, can be modelled using the position vector $\mathbf{r} = \begin{pmatrix} 8.66t \\ 5t - 5t^2 \end{pmatrix}$.

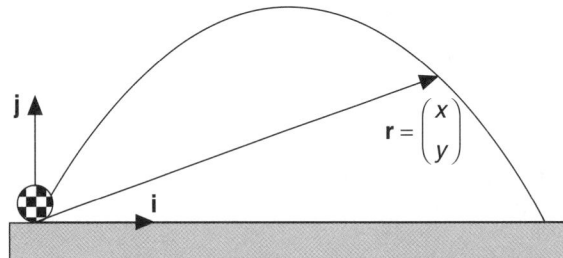

(a) List two of the assumptions that have been made in arriving at this solution. (2)
(b) Find how far away the ball is from where it was kicked when it first bounces. (4)
(c) Find the ball's speed just before its first bounce. (4)

UODLE

9   A racing car emerging from a bend reaches a straight stretch of road. The start of the straight stretch is the point O and there are two marker points, A and B, further down the road. The distance $OA = 64$ m and the distance $OB = 250$ m. The car passes O at time 0 s and, moving with constant acceleration, passes A and B at times 2 s and 5 s respectively. Find

(a) the acceleration of the car,
(b) the speed of the car at B. (7)

London Examinations

10   A particle P moves along the x-axis. It passes through the origin O at time $t = 0$ with speed $15\,m\,s^{-1}$ in the direction of x increasing. At time t seconds the acceleration of P in the direction of x increasing is $(6t - 18)\,m\,s^{-2}$.

(a) Find the values of t at which P is instantaneously at rest.
(b) Find the distance between the points at which P is instantaneously at rest. (12)

London Examinations

11 An aircraft lands with speed $u\,\mathrm{m\,s^{-1}}$ on the deck of an aircraft carrier. The aircraft is brought to rest in 2 seconds by an arrester gear and by the plane's own braking system. It is found that a good approximation to the speed, $v\,\mathrm{m\,s^{-1}}$, of the aircraft t seconds after touchdown is given by

$$v = u \cos \omega t - kt, \; 0 \leq t \leq 2$$

where ω and k are constants and $0 < \omega < 1$.

(a) Show that $k = \frac{1}{2}u \cos 2\omega$. (2)

(b) Show that, if $k = 2$ and $u = 32$, then $\omega \approx 0.7227$. (2)

(c) Using the values $k = 2$, $u = 32$, $\omega = 0.7227$,

    (i) tabulate to two decimal places the values of v for $t = 0.0, 0.4, 0.8, 1.2, 1.6, 2.0$; (2)

    (ii) use the trapezium rule with these six values of v to obtain an estimate of the distance travelled by the aircraft along the deck before it comes to rest. Give your answer to the nearest metre. (4)

                                                  NEAB

12 A small, delicate microchip which is initially at rest is to be moved by a robot arm so that it is placed **gently** onto a horizontal assembly bench. Two mathematical models have been proposed for the motion which will be programmed into the robot. In each model the unit of length is the centimetre and time is measured in seconds. The unit vectors **i** and **j** have directions which are horizontal and vertical respectively and the origin is the point O on the surface of the bench, as shown in the diagram.

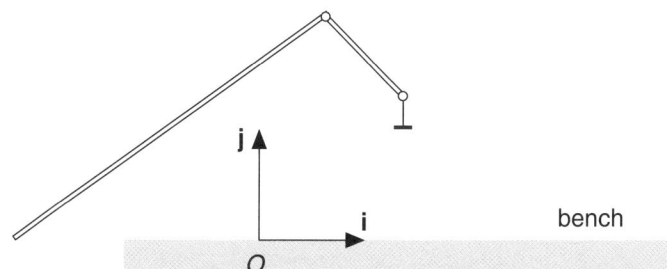

Model A for the position vector at time t of the microchip is $\mathbf{r}_A = 5t^2\mathbf{i} + (16 - 4t^2)\mathbf{j}$, $(t \geq 0)$.

(i) How far above the table is the microchip initially (i.e. when $t = 0$)? (1)

(ii) Show that this model predicts that the microchip reaches the table after 2 seconds and state the horizontal distance moved in this time. (3)

(iii) Calculate the predicted horizontal and vertical components of velocity when $t = 0$ and $t = 2$. (3)

Model B for the position vector at time t of the microchip is

$$\mathbf{r}_B = (15t^2 - 5t^3)\mathbf{i} + (16 - 24t^2 + 16t^3 - 3t^4)\mathbf{j}, \; (t \geq 0)$$

(iv) Show that model B predicts the same positions for the microchip at $t = 0$ and $t = 2$ as model A. (2)

(v) Calculate the predicted horizontal and vertical components of velocity for the microchip at $t = 0$ and $t = 2$ from model B and comment, with brief reasons, on which model you think describes the more suitable motion for the microchip. (5)

                                                  Oxford & Cambridge

3 Statics of a particle

Newton's Laws of Motion

The fundamental laws which govern the motion of bodies are **Newton's Laws of Motion** which may be summarised at this level as follows:

Law 1: A body remains at rest or continues to move in a straight line with uniform velocity unless it is acted upon by a resultant external force.

Law 2: Force = mass × acceleration or $\mathbf{F} = m\mathbf{a}$

Law 3: If body A exerts a force on body B then body B exerts an equal and opposite force on body A.

A **particle** is a point mass, i.e. a body which has no dimensions, which means we can ignore the rotational effect of any forces which are acting, since we cannot rotate a point.

A particle is an example of a **mathematical model**. Other examples are a **light** rod (a rod which has negligible mass), a **smooth** surface (a surface on which friction is negligible), an **inextensible** string (a string which does not stretch at all), a lamina (a body of negligible thickness) and a **rigid** body (a body which does not distort when subjected to the action of forces). Thus mathematical models are used to simplify real-world situations so that we have a framework on which to base our calculations. You should be aware of why certain models are adopted in certain situations and what their simplifying effect is.

A **particle** is in **equilibrium** is equivalent to saying that the **vector sum of the forces acting on it is zero**. (For a body which has dimensions we would also need to consider the rotational effects of the forces – we often **model** a large body by a particle so that we can ignore these rotational effects.)

If a particle is in equilibrium then it must be either at rest or moving with uniform velocity, by Newton's First Law. (Note that the converse is not true – it is possible for a particle to be at rest and yet not be in equilibrium.)

Solving problems

Suppose a particle P is in equilibrium under the action of **constant** forces $\mathbf{F}_1, \mathbf{F}_2, ..., \mathbf{F}_n$, then

$$\mathbf{F}_1 + \mathbf{F}_2 + ... + \mathbf{F}_n = 0$$

If the forces are given in **i-j** form, this is an easy equation to solve as we can just **add up the components and equate the sums to zero.**

If, however, the forces are given in magnitude-direction form, we need either to:

(i) **add the forces by drawing a vector polygon** which should be closed as the resultant is zero, and use trigonometry and/or Pythagoras, or

(ii) **draw a force diagram showing clearly the forces acting on P** and then equate the sum of the components or resolved parts of each of the forces, in two non-parallel (usually perpendicular) directions, to zero. This process is called **resolving in two directions**. We normally resolve horizontally and vertically or sometimes, if P is on an inclined plane, parallel and perpendicular to the plane. This will then give two equations which will need to be solved – in some problems one equation may suffice.

When drawing a force diagram, in addition to any external forces that are applied to the particle, the following types of force need to be considered: the **weight**, mg, acting vertically downwards, a **normal reaction**, R, if the particle is in contact with a surface, a **tension** if the particle is attached to a string and a **friction** force if it is in contact with a rough surface.

Friction

Friction *can* occur if a particle is in contact with a **rough** surface but will *only* act if there is a tendency for the particle to slide along the surface. The value of the friction force will always be **just sufficient** to prevent relative motion taking place, up to and including a maximum value, called **limiting friction**, given by μR, where μ is the coefficient of friction, a constant for the two given surfaces in contact (of course this maximum value μR may still not be sufficient to prevent sliding taking place). Thus $0 \leq F \leq \mu R$. If a stationary particle, which is in equilibrium, is just on the point of slipping it is said to be in **limiting equilibrium** and **in this case we can assume that** $F = \mu R$**.** If a particle is sliding along a rough surface (it may or may not be in equilibrium) then we can also assume that $F = \mu R$. It is in these two cases **only** that we can make this assumption.

Consider the following two examples:

(i) The force diagram below shows a particle, of mass m, at rest in equilibrium on a rough inclined plane of angle θ and coefficient of friction μ (where $\mu \geq \tan\theta$).

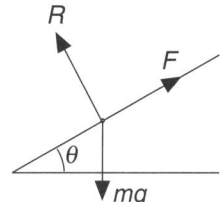

Resolving parallel to the plane: $F - mg\sin\theta = 0$

Resolving perpendicular to the plane : $R - mg\cos\theta = 0$

Notice that we have *not* used $F = \mu R$ since we are not told that we have limiting equilibrium.

(ii) The force diagram below shows a particle, of mass m, being dragged along a rough horizontal plane whose coefficient of friction is μ, at constant speed, using a light rope inclined at an angle θ to the plane.

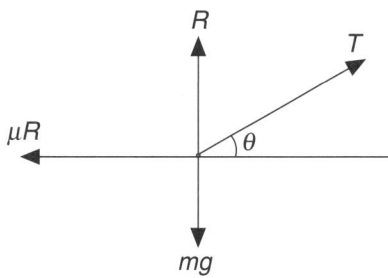

Resolving horizontally: $T\cos\theta - \mu R = 0$

Resolving vertically: $R + T\sin\theta - mg = 0$

Here $F = \mu R$ as the mass is moving.

If you need to revise this subject more thoroughly, see the relevant topics in the *Letts* **A-level** *Mathematics Study Guide.*

1

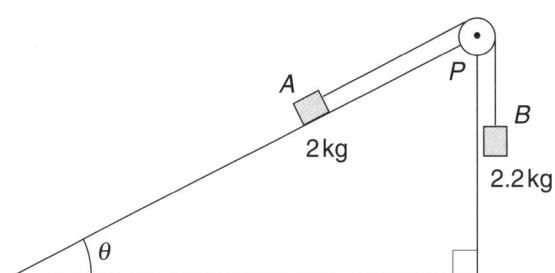

A parcel *A* of mass 2 kg rests on a rough slope inclined at an angle θ to the horizontal, where $\tan \theta = \frac{3}{4}$. A string is attached to *A* and passes over a small smooth pulley fixed at *P*. The other end of the string is attached to a weight *B* of mass 2.2 kg, which hangs freely, as shown in the diagram. The parcel *A* is in limiting equilibrium and about to slide up the slope. By modelling *A* and *B* as particles and the string as light and inextensible, find

(a) the normal contact force acting on *A*,

(b) the coefficient of friction between *A* and the slope. (10)

London Examinations

2   Three forces of 2, 3 and 4 newtons act on a particle which is in equilibrium, shown in the figure.

The 3 and 4 newton forces are replaced by one single force so that the particle remains in equilibrium. Give the magnitude and direction of this replacement force.

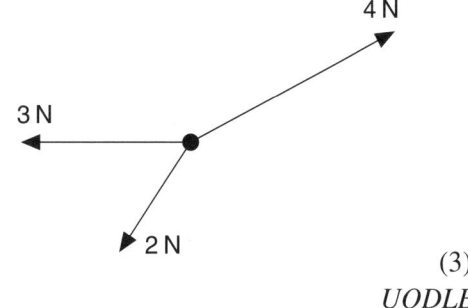

(3)

UODLE

3   A street light is hung over a busy main road using three wires. The tensions in the wires are T_1, T_2 and T_3, and **W** is the weight of the lamp. Unit vectors **i**, **j** and **k** are aligned so that the vector **i** points directly across the road and **k** is vertically upwards.

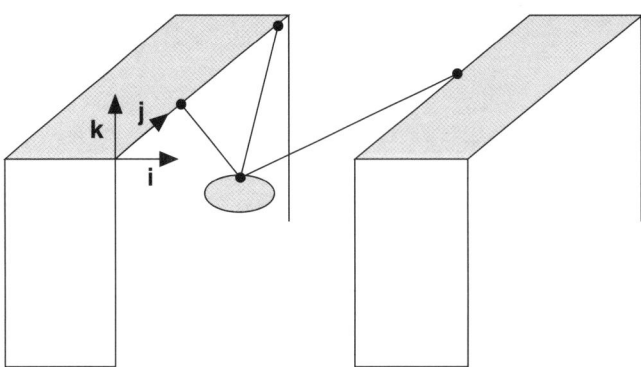

$T_1 = \begin{pmatrix} -40 \\ -40 \\ 10 \end{pmatrix}$ N, $T_2 = \begin{pmatrix} -40 \\ 40 \\ 10 \end{pmatrix}$ N and the mass of the street light is 5 kg. By considering the

equilibrium of the street light, find T_3. (4)

UODLE

4

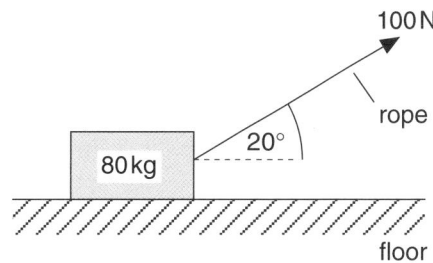

A box of mass 80 kg is to be pulled along a horizontal floor by means of a light rope. The rope is pulled with a force of 100 N and the rope is inclined at 20° to the horizontal, as shown in the diagram.

(i) Explain briefly why the box cannot be in equilibrium if the floor is smooth. (2)

In fact the floor is not smooth and the box is in equilibrium.

(ii) Draw a diagram showing all the external forces acting on the box. (2)

(iii) Calculate the frictional force between the box and the floor and also the normal reaction of the floor on the box, giving your answers correct to three significant figures. (5)

The maximum value of the frictional force between the box and the floor is 120 N and the box is now pulled along the floor with the rope always inclined at 20° to the horizontal.

(iv) Calculate the force with which the rope must be pulled for the box to move at a constant speed. Give your answer correct to three significant figures. (2)

(v) Calculate the acceleration of the box if the rope is pulled with a force of 140 N. (3)

Oxford & Cambridge

5 A washing line is made from a light rope stretched between two supports *A* and *C*, which are at the same height above the ground. The rope is lifted at *B* by means of a small, light pulley which is itself suspended by a light wire and the system is in equilibrium. The dimensions are shown in the diagram.

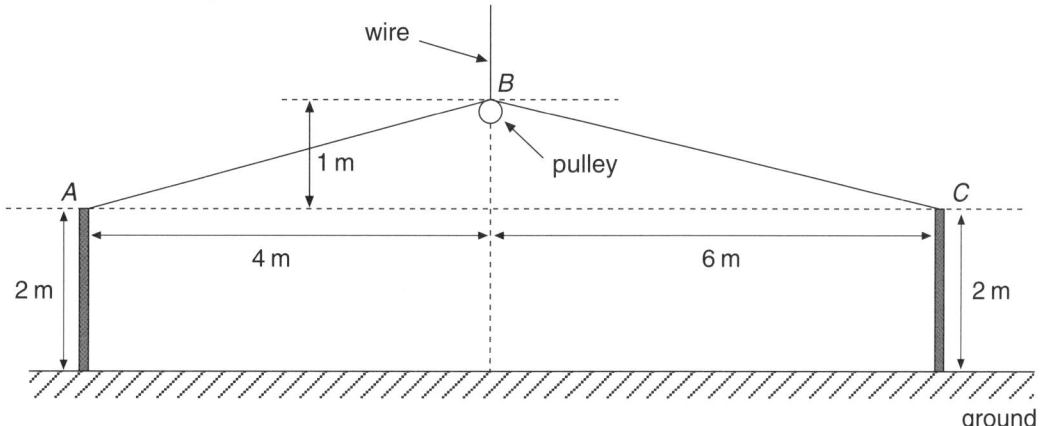

The tension in the rope section *AB* is 400 N and the pulley has smooth contact with the washing line at *B*.

(i) Calculate the angles that the rope sections *AB* and *BC* make with the horizontal. (1)

(ii) What is the tension in the rope section *BC*? (1)

(iii) Draw a diagram showing the forces acting on the pulley. (2)

(continued)

(iv) Calculate the horizontal and vertical components of the forces in the rope at *B*. Hence give the tension in the wire supporting the pulley as a vector in component form. You should state clearly which directions you have taken to be positive. (7)

A piece of washing is now fixed to the line at the mid-point of *BC*.

(v) Explain briefly why the rope section *BC* can no longer be straight. (3)

Oxford & Cambridge

6

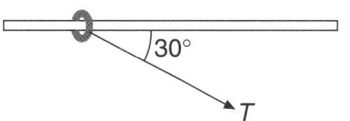

A heavy ring of mass 5 kg is threaded on a fixed rough horizontal rod. The coefficient of friction between the ring and the rod is $\frac{1}{2}$. A light string is attached to the ring and pulled downwards with a force acting at a constant angle of $30°$ to the horizontal (see diagram). The magnitude of the force is *T* newtons, and is gradually increased from zero. Find the value of *T* that is just sufficient to make the equilibrium limiting. (6)

UCLES

7 A small smooth ring *R* of mass 0.1 kg is threaded on a light string. The ends of the string are fastened to two fixed points *A* and *B*. The ring hangs in equilibrium with the part *AR* of the string inclined at $40°$ to the horizontal, as shown in the diagram. Show that the part *RB* of the string is also inclined at $40°$ to the horizontal, and find the tension in the string. (4)

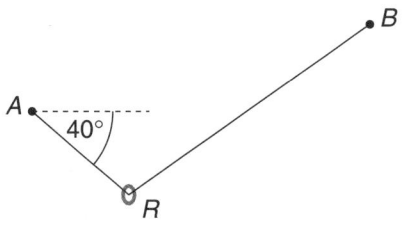

UCLES

8 Susie uses a strap to pull her suitcase, at a constant speed in a straight line, along the horizontal floor of an airport departure lounge. The strap is inclined at $50°$ to the horizontal and the frictional force exerted on the case by the floor has magnitude 20 N. Modelling the suitcase as a particle, find the tension in the strap. (3)

UCLES

9

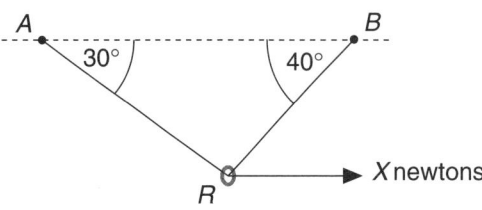

A small smooth ring *R*, of mass 0.3 kg, is threaded on a light inextensible string whose ends are attached to two fixed points *A* and *B* which are at the same horizontal level. A force of magnitude *X* newtons is applied to the ring in a direction parallel to *AB*, as shown in the diagram. When the ring is in equilibrium with both parts of the string taut, angle $BAR = 30°$ and angle $ABR = 40°$. Find the tension in the string and the value of *X*. (6)

UCLES

REVISION SUMMARY

Suppose a particle P of mass m is moving in a straight line with acceleration $\mathbf{a}$, under the action of **constant** forces $\mathbf{F}_1$, $\mathbf{F}_2$,..., $\mathbf{F}_n$, then by Newton's Second Law,

$$\mathbf{F}_1 + \mathbf{F}_2 + ... + \mathbf{F}_n = m\mathbf{a}$$

If the forces are given in **i-j** form then we can just add the components of **i** and **j** to find the resultant force. If, however, the forces are given in magnitude-direction form then, in a similar way to the statics case, we need to draw a clear force diagram, showing the acceleration as well, and **resolve the forces and the acceleration** in two non-parallel directions.

In practice we usually **resolve in the direction of the acceleration**, so that the RHS of the equation is ma, and also in a direction which is **perpendicular to the acceleration**, so that the RHS of the equation is zero. This gives more straightforward equations to solve. The techniques used are thus very similar to those used for statics, except that when resolving in the direction of the acceleration the RHS of the equation will be ma instead of zero.

Connected particles

If a system consists of two particles which are connected by, for example, a light and inextensible string, and the particles are moving in the same straight line, then we can obtain equations by **resolving separately for each particle and for the system as a whole**. *If the particles are not moving in the same straight line*, then the whole system equation is *not* available.

The **work done** by a constant force $\mathbf{F}$ when it moves its point of application through a displacement $\mathbf{d}$ is given by $\mathbf{F.d} = Fd\cos\theta$. Work is therefore a scalar and is measured in **joules.**

Doing work on a particle changes its **energy**, the unit of which is also the joule. The **kinetic energy** (K.E.) of a particle is energy due to its motion and is given by $\frac{1}{2}mv^2$. Its **potential energy** (P.E.) is energy due to its position and is given by mgh. Both are scalars.

Work done against a resistance = overall loss of energy = all losses – all gains

Work done on a particle = overall gain in energy = all gains – all losses

If no work is done then All losses = All gains. This is known as the **Conservation of Energy Principle**.

Power is the rate of doing work and is measured in **watts**. For a moving vehicle, at any instant, the power being developed by its engine = driving force of engine × speed, at that instant, i.e. $P = Fv$.

Momentum = $m\mathbf{v}$ is a vector measured in N s.

The **impulse** of a constant force $\mathbf{F}$ which acts for a time t is given by $\mathbf{F}t$, is also a vector and is also measured in N s.

Impulse = change in momentum, i.e. $\mathbf{I} = m\mathbf{v} - m\mathbf{u}$. In any collision between two particles, each particle receives an equal and opposite impulse, i.e. the gain in momentum of one particle = loss in momentum of the other, thus the total momentum of the system is unchanged. This is known as the **Conservation of Momentum Principle**. Note that, although momentum is always conserved in a collision, usually there is a loss of kinetic energy due to energy being lost to sound and/or heat.

If you need to revise this subject more thoroughly, see the relevant topics in the *Letts A-level Mathematics Study Guide*.

1 A woman of mass 60 kg runs along a horizontal track at a constant speed of $4\,\text{m s}^{-1}$. In order to overcome air resistance, she works at a constant rate of 120 W.

(a) Find the magnitude of the air resistance which she experiences.

She now comes to a hill inclined at an angle α to the horizontal where $\sin \alpha = \frac{1}{15}$.

To allow for the hill, she reduces her speed to $3\,\text{m s}^{-1}$ and maintains this constant speed as she runs up the slope. In a preliminary model of this situation, the air resistance is modelled as having the constant value obtained in (a) whatever the speed of the woman.

(b) Estimate the rate at which the woman has to work against the external forces in order to run up the hill.

In a more refined model, the air resistance experienced by the woman is taken as proportional to the square of her speed.

(c) Use your answer to (a) to obtain a revised estimate of the air resistance experienced by the woman when running at $3\,\text{m s}^{-1}$.

(d) Find a revised estimate of the rate at which the woman has to work against the external forces as she runs up the hill. (11)

London Examinations

2 A laundry basket of mass 3 kg is being pulled along a rough horizontal floor by a light rope inclined upward at an angle of $30°$ to the floor. The tension in the rope is 8 N. Considering the laundry basket as a particle, calculate the magnitude of the normal component of the resultant force exerted on the laundry basket by the floor. (3)

Given that the acceleration of the laundry basket is $0.2\,\text{m s}^{-2}$, find the coefficient of friction between the laundry basket and the floor. (4)

UCLES

3 The diagram shows a man of mass 70 kg standing in a lift and carrying a suitcase of mass 8 kg in his hand. The magnitude of the contact force between the man and the floor of the lift is R newtons and that between the suitcase and the man's hand is S newtons. Take $g = 10\,\text{m s}^{-2}$.

(a) (i) Show in a diagram the two forces acting on the suitcase. (1)

(ii) Show in a further diagram the three forces acting on the man. (1)

(b) Calculate the values of R and S when the upward acceleration of the lift is $2\,\text{m s}^{-2}$. (4)

NEAB

4 Pat and Nicholas are controlling the movement of a canal barge by means of long ropes attached to each end. The tension in the ropes may be assumed to be horizontal and parallel to the line and direction of motion of the barge, as shown in the diagrams.

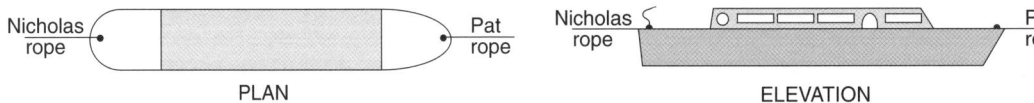

Nicholas rope · · Pat rope   PLAN   Nicholas rope · · Pat rope   ELEVATION

The mass of the barge is 12 tonnes and the total resistance to forward motion may be taken to be 250 N at all times.

Initially Pat pulls the barge forwards from rest with a force of 400 N and Nicholas leaves his rope slack.

(i)   Write down the equation of motion for the barge and hence calculate its acceleration. (4)

Pat continues to pull with the same force until the barge has moved 10 m.

(ii) What is the speed of the barge at this time and for what length of time did Pat pull?   (4)

Pat now lets her rope go slack and Nicholas brings the barge to rest by pulling with a constant force of 150 N.

(iii) Calculate

  (A) how long it takes the barge to come to rest,

  (B) the total distance travelled by the barge from when it first moved,

  (C) the total time taken for the motion.   (6)

Oxford & Cambridge

5 A car of mass 800 kg is travelling at a steady speed of $25\,\mathrm{m\,s^{-1}}$ along a straight, level road. The car engine is working at the rate of 40 kW.

(i)   Calculate the resistance to motion.   (2)

The car reaches a slope with an angle of $\arcsin\frac{1}{10}$ which it climbs at the same speed against the same resistance.

(ii) What extra power must the car engine produce?   (5)

Whilst the car is climbing the slope at a speed of $25\,\mathrm{m\,s^{-1}}$ the power is suddenly removed and the car slows down and comes to rest. Whilst the car is slowing down the resistance to motion may be taken to have a constant value of 900 N.

(iii) How far along the slope does the car travel whilst slowing down to come to rest?   (7)

Oxford & Cambridge

6 (a) A batsman strikes a cricket ball of mass 0.15 kg at right angles to the bat so that the direction of the ball is reversed. The ball is travelling horizontally at $25\,\mathrm{m\,s^{-1}}$ just before it is struck by the bat, and leaves the bat with a speed of $40\,\mathrm{m\,s^{-1}}$ as shown in the figure.

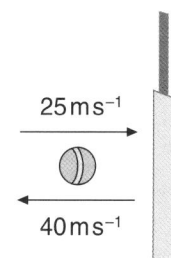
$25\,\mathrm{ms^{-1}}$

$40\,\mathrm{ms^{-1}}$

  (i)   What is the magnitude of the impulse exerted on the ball by the bat?   (4)

  (ii) If the bat and ball are in contact for 0.01 s, what is the average force exerted on the ball by the bat?   (3)

(b) During a game of bowls, a bowl of mass 1.5 kg moves with a speed of $1\,\mathrm{m\,s^{-1}}$ on a smooth, horizontal green. It strikes directly a stationary bowl of mass 1.2 kg. The 1.2 kg bowl then begins to move with a speed of $0.8\,\mathrm{m\,s^{-1}}$. What is the speed of the 1.5 kg bowl immediately after the collision?   (5)

NICCEA

7 (In this question take g as $10\,\text{m s}^{-2}$.)

After an unexpected snow fall some children take tin trays out to a local hillside and slide down the hillside sitting on the tin trays.

(a) List three factors that would affect the amount of friction between the tray and the snow. (3)

The hill is inclined at $30°$ to the horizontal.

(b) (i) Show that if friction is assumed to be zero then the acceleration of a child on a tray will be $5\,\text{m s}^{-2}$. (2)

(ii) Find the time it takes for the child to travel $20\,\text{m}$ down the slope, if the child starts at rest. (2)

(c) In reality the child is given a push so that the initial speed is $2\,\text{m s}^{-1}$ and it then takes 4 seconds to travel the $20\,\text{m}$. Find the actual acceleration of the child. (2)

(d) What is the coefficient of friction between the tray and the snow? (3)

(e) State one other factor that has not been taken into account in the problem. (1)

AEB

8 A block of wood, of mass $2\,\text{kg}$, is at rest on a smooth horizontal table. A bullet, of mass $0.1\,\text{kg}$, moving horizontally at a speed of $420\,\text{m s}^{-1}$, strikes the block and becomes embedded in it. Find the speed of the block after the impact. (3)

UCLES

9 A car has an engine of maximum power $15\,\text{kW}$. Calculate the force resisting the motion of the car when it is travelling at its maximum speed of $120\,\text{km h}^{-1}$ on a level road. (3)

Assuming an unchanged resistance, and taking the mass of the car to be $800\,\text{kg}$, calculate, in m s^{-2}, the maximum acceleration of the car when it is travelling at $60\,\text{km h}^{-1}$ on a level road. (4)

UCLES

10

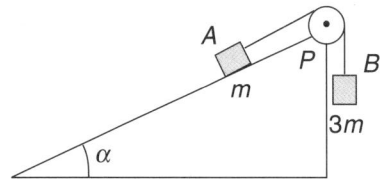

A block A of mass m can move on the rough face of a wedge inclined at an angle α to the horizontal, where $\sin \alpha = 0.6$. The wedge is fixed to horizontal ground. A rope is attached to A and passes over a small smooth pulley fixed at P. The other end of the rope is attached to a block B of mass $3m$ which hangs freely vertically below P, as shown in the figure. The blocks are modelled as particles and the rope is assumed to be light and inextensible and sufficiently long for A not to hit the pulley in the subsequent motion. The coefficient of friction between A and the plane is $\frac{3}{4}$. The system is released from rest with the rope taut and with B initially $1\,\text{m}$ above the ground.

(a) Find the acceleration of A in the period before B hits the ground.

When B hits the ground, the rope becomes slack and A continues to move up the slope.

(b) Calculate the total distance moved by A before it first comes to rest. (17)

London Examinations

11 A truck *A* of mass 900 kg moving on a straight horizontal railway line with speed 7 m s⁻¹ collides with a second truck *B* of mass 500 kg which is stationary. At the collision the trucks are automatically coupled and move off together. By modelling the trucks as particles,

(a) show that the speed of the trucks immediately after the collision is 4.5 m s⁻¹,

(b) find the magnitude of the impulse exerted on *B* by *A* in the collision,

(c) find the kinetic energy lost in the collision.

After the collision, the trucks continue to move with a constant speed of 4.5 m s⁻¹. They crash into some buffers which bring the trucks to rest. The buffers provide a total constant force of 95 000 N as they are compressed.

(d) Find the total distance by which the buffers are compressed in bringing the trucks to rest. (14)

London Examinations

12 A small block is pulled along a rough horizontal surface at a constant speed of 2 m s⁻¹ by a constant force. This force has magnitude 25 N and acts at an angle of 30° to the horizontal. Calculate the work done by the force in 10 seconds. (3)

UCLES

13 A child kicks his toy boat, which has mass 1 kg, across the horizontal path at the side of the harbour. Initially the boat is at the point *P*. The boat moves in a straight line which is at right angles to one edge of the path, reaching this edge at the point *Q*, where *PQ* = 2 m. The speeds of the boat at *P* and *Q* are 10 m s⁻¹ and 8 m s⁻¹ respectively. Modelling the boat as a particle, and assuming that air resistance may be ignored, find the coefficient of friction between the boat and the path. (4)

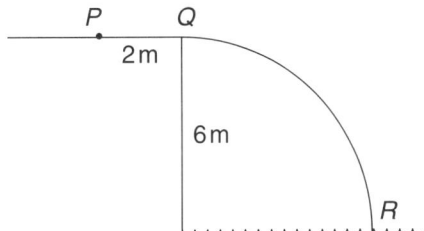

The path is at a height of 6 m above the water. After the boat leaves the path it moves freely under gravity and hits the water at the point *R* (see diagram). Making the same modelling assumptions as before, find the horizontal distance of *R* from *Q*. (3)

The point *S* on the boat's trajectory is the point at which the boat is moving in a direction inclined at 45° to the horizontal. Find the height of *S* above the water. (4)

UCLES

5 *Further momentum*

In Unit 4 the conservation of momentum principle was introduced for any collision between two particles. This enables some straightforward impact questions to be solved. A further equation, called **Newton's Experimental Law**, or the **Restitution Equation**, enables calculation of the velocities after impact. This law states that the relative velocity along the common normal after impact is in a constant ratio to the relative velocity before impact and is reversed in direction. The ratio, e, is the coefficient of restitution. This is a measure of the elasticity of the bodies in collision and $0 \leq e \leq 1$. If $e = 0$ the collision is inelastic. If $e = 1$ then the collision is perfectly elastic.

Direct impact of two particles

A diagram showing the magnitudes and directions of the given velocities should be drawn, and two equations written down.

The Conservation of Linear Momentum (C.L.M.) gives $m_1 u_1 + m_2 u_2 = m_1 v_1 + m_2 v_2$.

Newton's Experimental Law (N.E.L.) gives $e(u_1 - u_2) = v_2 - v_1$.

These equations are then solved using techniques for solving simultaneous equations.

If any velocity is in the opposite direction then its sign must be taken as negative.

Loss of kinetic energy: provided that $e \neq 1$ there is always loss of kinetic energy at an impact, and in the above example this is $\frac{1}{2} m_1 u_1^2 + \frac{1}{2} m_2 u_2^2 - \frac{1}{2} m_1 v_1^2 - \frac{1}{2} m_2 v_2^2$.

Normal impact of a particle on a surface

A particle of mass m moves with speed $u\, \text{m s}^{-1}$ at right angles to a fixed surface.

It rebounds with speed $v\, \text{m s}^{-1}$.

$$e = \frac{\text{speed of separation}}{\text{speed of approach}} \Rightarrow v = eu$$

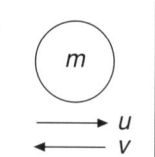

where e is the coefficient of restitution for the collision. The change of momentum is absorbed by the wall and the impulse exerted on the particle has magnitude $m(u + eu)$.

Oblique impact of a particle on a surface

A particle of mass m moves with speed $u\, \text{m s}^{-1}$ and hits a wall at an angle θ to the wall. It rebounds with a speed $v\, \text{m s}^{-1}$ at an angle ϕ to the wall, as shown in the figure.

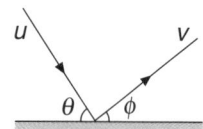

C.L.M. along the wall: $\qquad u \cos \theta = v \cos \phi$

N.E.L. perpendicular to the wall: $eu \sin \theta = v \sin \phi$

From these two equations, $\tan \phi = e \tan \theta$ and $v^2 = u^2 (\cos^2 \theta + e^2 \sin^2 \theta)$.

If you need to revise this subject more thoroughly, see the relevant topics in the *Letts* A-level *Mathematics Study Guide.*

Oblique impact of two particles

The velocity component at right angles to the line of centres is unchanged. Along the line of centres the two equations arising from C.L.M. and N.E.L. are used. So

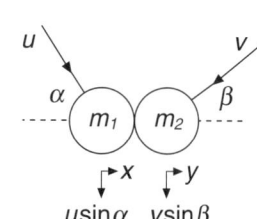

$m_1 u \cos \alpha - m_2 v \cos \beta = m_1 x + m_2 y$

$e(u \cos \alpha + v \cos \beta) = y - x$

Further collisions: when there is more than one collision, each is considered separately using the two equations as before.

1 Two identical particles are moving in the same straight line. Immediately before they collide, they are moving with speeds $1\,\mathrm{m\,s^{-1}}$ and $2\,\mathrm{m\,s^{-1}}$ in the same direction. Given that the coefficient of restitution is 0.8, calculate the speeds of the particles after the collision. (4)

UCLES

2 A small smooth sphere is dropped from a point at a height of $0.6\,\mathrm{m}$ above a smooth horizontal floor. The sphere falls vertically, strikes the floor and bounces to a height of $0.15\,\mathrm{m}$ above the floor.

(a) Find the speed of the sphere when it first hits the floor. (1)

(b) Find the coefficient of restitution between the sphere and the floor. (3)

WJEC

3

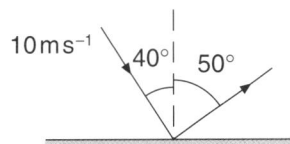

A table tennis ball of mass $2.5\,\mathrm{g}$ strikes the smooth surface of a horizontal table. Just before the impact the speed of the ball is $10\,\mathrm{m\,s^{-1}}$, and its direction of motion makes an angle of 40° with the vertical. Just after the impact the direction of motion makes an angle of 50° with the vertical (see diagram). Calculate

(i) the speed of the ball immediately after the impact, (3)

(ii) the magnitude of the impulse of the table on the ball. (4)

UCLES

4 A uniform smooth sphere P, of mass $3m$, is moving in a straight line with speed u on a smooth horizontal table. Another uniform smooth sphere Q, of mass m and having the same radius as P, is moving with speed $2u$ in the same straight line as P but in the opposite direction to P. The sphere P collides with the sphere Q directly. The velocities of P and Q after the collision are v and w respectively, measured in the direction of motion of P before the collision. The coefficient of restitution between P and Q is e.

(a) Find an expression for v in terms of u and e.

(b) Show that, if the direction of motion of P is changed by the collision, then $e > \frac{1}{3}$.

(c) Find an expression for w in terms of u and e.

(d) Show that, as e varies, w can never exceed $\dfrac{5u}{2}$.

Following the collision with P, the sphere Q then collides with and rebounds from a vertical wall which is perpendicular to the direction of motion of Q. The coefficient of restitution between Q and the wall is e'.

Given that $e = \frac{5}{9}$, and that P and Q collide again in the subsequent motion,

(e) show that $e' > \frac{1}{9}$. (18)

London Examinations

5

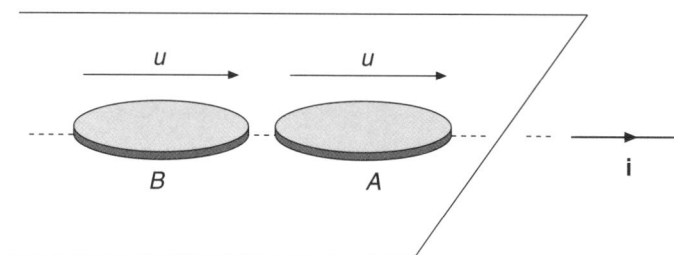

Two identical circular discs, each of mass M, are sliding on a smooth horizontal table with a constant velocity of u **i** m s^{-1} with their centres on the line with direction **i**, as shown in the diagram.

An insect of mass $\frac{1}{4}M$ drops from rest straight down onto the centre of disc A and clings to the point where it lands.

(i) Calculate the new velocity of disc A. (2)

The insect now jumps off the disc with a velocity of u **i** m s^{-1} (with respect to the table).

(ii) Calculate the new velocity of disc A. (3)

Disc A is now struck directly by disc B in an impact where the speed of separation is half the speed of approach.

(iii) Calculate the velocities of the two discs after the impact and also the impulse given to disc B by disc A. (7)

While the insect was on disc A, it started to walk from the centre with constant velocity in the direction **i** and stopped when it reached the edge of the disc.

(iv) Describe the motion of the disc. You may assume that there was no impact of the discs during the walk. (2)

Oxford & Cambridge

6

Not to scale

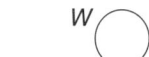

In a game of snooker, the ball W is resting against the cushion at the edge of the table (see diagram). The ball R is stationary, with its centre 325 mm from the same cushion. Each ball has a radius of 25 mm. The centres of W and R are 500 mm apart. A player decides to strike W in such a way that it will hit R, and cause R to move with a speed of 100 mm s^{-1} parallel to the cushion. The situation can be modelled as a collision between smooth uniform spheres of equal mass.

(i) Copy the diagram, and show on your diagram the position of W at the instant of its collision with R. (1)

(ii) Prove that, before the collision, the path of W must make an angle with the cushion of $40.6°$ correct to 3 significant figures. (3)

(iii) Given that the coefficient of restitution is 0.9, calculate the required speed of W just before the collision. (5)

UCLES

In previous summaries we have considered the linear motion of a particle which is subjected to constant forces. We shall now consider what happens when these forces are **varying**. By Newton's second Law, $\mathbf{F} = m\mathbf{a}$, and so it follows that, provided that m is a constant, if $\mathbf{F}$ is a variable then $\mathbf{a}$ will be, and so we shall have to use a derivative form for the acceleration, either $\dfrac{dv}{dt}$ or $v\dfrac{dv}{dx}$. Note that for motion in a straight line a vector can be replaced by a positive or negative scalar.

REVISION SUMMARY

Solving problems

Draw a simple diagram showing the forces acting and the acceleration – note that a must be put on your diagram *in the positive x-direction*. Resolve in the direction of a, and then replace a by one of the two derivative forms – **use $\dfrac{dv}{dt}$ if the force is a function of time** and $v\dfrac{dv}{dx}$ **if the force is a function of distance**. **If the force is a function of velocity,** which of the two forms you will need to use will be determined by the question. You will then have a first order differential equation which can be solved by separating the variables and integrating both sides of the equation. *The arbitrary constant of integration can be determined by using the initial conditions.*

Impulse and work done by a variable force: applying $F = ma$ leads to

$$F = m\,\frac{dv}{dt} \qquad \text{or} \qquad F = m\,\frac{v\,dv}{dx}$$

$$\int_{t_1}^{t_2} F\,dt = m\int_{u}^{v} dv \qquad\qquad \int_{x_1}^{x_2} F\,dx = m\int_{u}^{v} v\,dv$$

$$\int_{t_1}^{t_2} F\,dt = mv - mu \qquad\qquad \int_{x_1}^{x_2} F\,dx = \tfrac{1}{2}mv^2 - \tfrac{1}{2}mu^2$$

$\displaystyle\int_{t_1}^{t_2} F\,dt$ is defined to be the **impulse** of the variable force F.

$\displaystyle\int_{x_1}^{x_2} F\,dx$ is defined to be the **work done** by the variable force F.

Thus impulse = change in momentum

Thus work done = change in KE

Note that for a *constant* force F,

Note that for a *constant* force F,

$$\text{Impulse} = \int_{t_1}^{t_2} F\,dt = F(t_2 - t_1) \qquad\qquad \text{Work done} = \int_{x_1}^{x_2} F\,dx = F(x_2 - x_1)$$

If you need to revise this subject more thoroughly, see the relevant topics in the *Letts* A-level Mathematics Study Guide.

Newton's Law of Gravitation

A particular example of a force which is a function of distance is gravitational attraction which is given by $F = \dfrac{GM_1 M_2}{d^2}$. This gives the magnitude F of the gravitational force of attraction between two bodies of masses M_1 and M_2, which are a distance d apart. G is a constant known as the **constant of gravitation**. This result is a statement of **Newton's Law of Gravitation**.

For a particle of mass m on the surface of the Earth, $F = mg$. Hence $\dfrac{GMm}{R^2} = mg$, where M is the mass of the Earth and R is the radius of the Earth.

Thus $G = \dfrac{R^2 g}{M}$. The units of G are $kg^{-1}\,m^3\,s^{-2}$ and its value is 6.67×10^{-11} (approximately).

1 The acceleration of a particle is given by $(10 - 0.5\,v^2)\,\mathrm{m\,s^{-2}}$, where v is its speed in metres per second. Initially the particle is at rest. Find how far the particle has travelled when its speed has reached $2\,\mathrm{m\,s^{-1}}$. (10)

NICCEA

2 A cyclist is travelling along a road.

(a) What can be deduced about the resultant force on the cycle and cyclist if they are travelling at top speed (i) along a straight road, (ii) along a winding road? (3)

A cyclist whose maximum rate of working is 600 W can reach a top speed of $10\,\mathrm{m\,s^{-1}}$ on a level road. The combined mass of the cycle and cyclist is 90 kg.

(b) By assuming that the resistance forces of the cycle and cyclist are proportional to their speed, find a simple model for the total resistance force. (3)

(c) By assuming that the forward force on the cyclist is constant, show that
$$\frac{\mathrm{d}v}{\mathrm{d}t} = -\frac{10 - v}{15}$$
where $v\,\mathrm{m\,s^{-1}}$ is the speed of the cyclist at time t seconds. (3)

(d) Find an expression for the speed of the cyclist in terms of time, if he starts at rest. (5)

(e) Criticise your model for the resistance on the cyclist. (1)

AEB

3 Assume that the gravitational attraction of the Earth on an object of mass m at a distance r from the centre of the Earth is $\dfrac{km}{r^2}$, where k is a positive constant. A rocket is launched from the Earth's surface, and it travels vertically upwards. When the fuel is exhausted, the distance of the rocket from the centre of the Earth is a and the speed of the rocket is u. Some time later, the distance of the rocket from the centre of the Earth is x and the speed of the rocket is v. Neglecting any forces other than the gravitational attraction of the Earth, find an expression for v. (8)

Deduce that, if $u^2 \geq \dfrac{2k}{a}$, the rocket will never fall back to Earth. (2)

UCLES

4 (a) In a simple model to find the time taken for a car to brake from a speed of $u\,\mathrm{m\,s^{-1}}$ to rest, it is assumed that the only horizontal force acting on the car is the braking force. This force is taken to be proportional to the weight mg of the car, where m is its mass in kilograms.

  (i) Show that the acceleration of the car when braking is given by $\dfrac{\mathrm{d}v}{\mathrm{d}t} = -kg$ where k is a positive constant and v is the speed of the car t seconds after braking starts. (2)

  (ii) Find v in terms of u, k, g and t. (2)

(b) A refinement of the model takes air resistance, R, into account. Assume this is proportional to the speed of the car and thus take $R = bmv$, where b is a positive constant.

  (i) Show that the refined model gives the equation $\dfrac{\mathrm{d}v}{\mathrm{d}t} = -bv - kg$. (2)

  (ii) Hence show that $t = \dfrac{1}{b}\ln\left(\dfrac{bu + kg}{bv + kg}\right)$. (4)

(c) Taking $g = 10\,\mathrm{m\,s^{-2}}$, $k = 0.7$ and $b = 0.2$, find to the nearest 0.01 second the difference in the times given by the two models for the car to brake from $30\,\mathrm{m\,s^{-1}}$ to rest. (3)

NEAB

Moments, centres of mass and equilibrium 7

The **moment** of a force is a measure of its *turning effect*.

The moment of a force F about an axis through a given point P is defined as:

$$\text{Moment} = F \times x$$

where x is the perpendicular distance from the point P to the line of action of the force F.

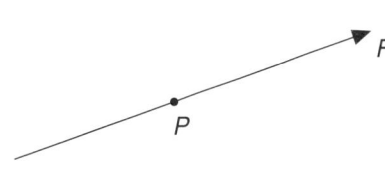

The units of moment are N m (newton metres), and moments can be clockwise or anti-clockwise in sense.

When a force acts *through* the point P, its moment about P is zero.

When the line of action of the force F meets a fixed line through P at a point A, where the distance AP is d and where the angle between the direction of the force and the line is θ:

$$\text{The moment of } F \text{ about } P \text{ is } Fd \sin \theta.$$

This may be obtained by finding the perpendicular distance, see (a), or by resolving force F into components, see (b).

(a)

Moment $= Fx = Fd \sin \theta$

(b)

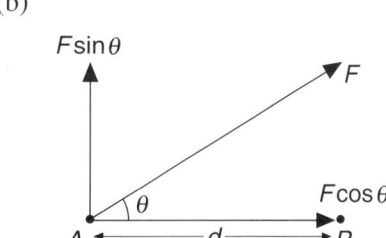

$F \cos \theta$ acts through P, so its moment $= 0$

Moment $=$ moment of $F \sin \theta = Fd \sin \theta$

The combined turning effect of a number of **coplanar** forces about an axis is given by the algebraic sum of the clockwise and anti-clockwise moments, where one sense is taken as positive and the other as negative.

Moments are used in centres of mass problems and in equilibrium problems which involve a rigid body rather than a particle.

Centres of mass

The **centre of mass** of a rigid body is *the point at which the weight acts*, in a uniform gravitational field.

For a uniform rigid body, the weight is evenly distributed and so the centre of mass must lie on any axis of symmetry.

Given that the centre of mass of n masses m_1, m_2, m_3, ..., m_n at the points (x_1, y_1), ... (x_n, y_n) lies at $(\bar{x}, \bar{y})$ then

$$\bar{x} = \frac{\sum m_i x_i}{\sum m_i}, \quad \bar{y} = \frac{\sum m_i y_i}{\sum m_i}$$

Some standard results

Body	Centre of mass
Uniform rod	Midpoint of rod
Uniform circular disc	Centre of circle
Uniform triangular lamina	Intersection of medians, i.e. $\frac{2}{3}$ way from vertex to midpoint of opposite side of the triangle
Uniform rectangular lamina	Point of intersection of lines joining midpoints of opposite sides
Uniform sphere	Centre of sphere
Uniform cylinder	Midpoint of axis of symmetry

The position of the centres of mass of these common bodies can be found by considering symmetries. Questions are likely to involve one or more of these bodies combined together, or one body *removed* from another. Moments are used to find the position of the centre of mass of the resulting body, and it is helpful to draw a table showing the masses and the positions of the centres of mass of the separate parts of the body together with the total mass of the combined body.

Calculus methods

Calculus can be used to determine the position of the centre of mass of a lamina bounded by a curve, with equation $y = f(x)$, as shown below.

Calculus methods are not needed for all syllabuses; check to see whether you need to read this section.

The formulae used are:

$$\overline{x} = \frac{\displaystyle\int_a^b x\,f(x)\,dx}{\displaystyle\int_a^b f(x)\,dx}, \quad \overline{y} = \frac{\displaystyle\frac{1}{2}\int_a^b [f(x)]^2\,dx}{\displaystyle\int_a^b f(x)\,dx}$$

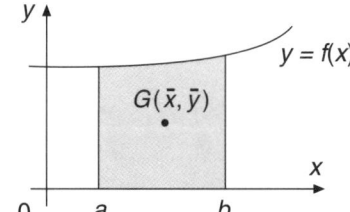

Solids of revolution

The centre of mass of any solid body will lie on any axis or plane of symmetry. For a solid of revolution an axis of symmetry will be one of the coordinate axes.

The diagram below shows a solid of revolution obtained by rotating the curve $y = f(x)$ through $360°$, about the x-axis. Its centre of mass lies on the x-axis.

Also the x coordinate of its centre of mass is given by the formula:

$$\overline{x} = \frac{\displaystyle\int_a^b \pi[f(x)]^2 x\,dx}{\displaystyle\int_a^b \pi[f(x)]^2\,dx}$$

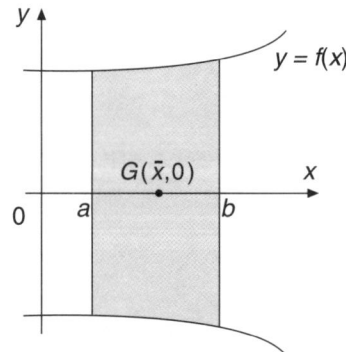

Further standard results

Rigid body	*Centre of mass*
Solid hemisphere, radius r	$\frac{3}{8}r$ from centre
Hemispherical shell, radius r	$\frac{1}{2}r$ from centre
Solid right circular cone, height h	$\frac{3}{4}h$ from vertex
Sector of circle, radius r, angle at centre 2θ	$\dfrac{2r\sin\theta}{3\theta}$ from centre
Semicircular lamina, radius r	$\dfrac{4r}{3\pi}$ from centre
Circular arc, radius r, angle at centre 2θ	$\dfrac{r\sin\theta}{\theta}$ from centre
Semicircular arc, radius r	$\dfrac{2r}{\pi}$ from centre

Positions of equilibrium

When a rigid body is freely suspended and it hangs in equilibrium, its centre of mass is vertically below the point of suspension.

When a rigid body rests in equilibrium on an inclined plane, its centre of mass lies vertically above a point of contact between the body and the plane.

Most centre of mass questions require you to use these facts together with simple trigonometry to find a position of equilibrium.

Conditions for equilibrium

For a rigid body to be in a state of equilibrium under the action of a system of coplanar forces two things have to be true:

(i) The sum of the components of the forces in any two directions must be zero.

(ii) The algebraic sum of the moments of the forces about any point in their plane must be zero.

It is usual to resolve horizontally and vertically, or along an inclined plane and perpendicular to the plane.

Moments are usually taken about a point through which several unknown forces are acting, in order to eliminate these forces from the equation. Since the forces act through the chosen point, their moments are zero about it.

If a rigid body remains in equilibrium under the action of three coplanar forces, then either all three forces are parallel, or their lines of action meet at a single point (they are concurrent).

If you need to revise this subject more thoroughly, see the relevant topics in the *Letts* A-level Mathematics Study Guide.

7 Moments, centres of mass and equilibrium

1

A rigid goal frame of mass 100 kg consists of two identical posts, P_1 and P_2, and a uniform crossbar. When the frame is placed in suitably constructed holes in a playing field, the posts are vertical and the crossbar is horizontal. The contact forces on the frame, which are assumed to act vertically at the base of the holes, have magnitudes R_1 newtons and R_2 newtons, as indicated in the diagram. State the value of R_1. (2)

A goalkeeper of mass 75 kg hangs by one hand, without moving, from a point of the crossbar 2 m from the corner where P_1 meets the crossbar. Given that the length of the crossbar is 7 m, calculate the new values of R_1 and R_2. (4)

UCLES

2 The diagram shows a light rod AB of length $4a$ rigidly joined at B to a light rod BC of length $2a$ so that the rods are perpendicular to each other and in the same vertical plane. The centre O of AB is fixed and the rods can rotate freely about O in a vertical plane. A particle of mass $4m$ is attached at A and a particle of mass m is attached at C. The system rests in equilibrium with AB inclined at an acute angle θ to the vertical as shown. By taking moments about O, find the value of θ.

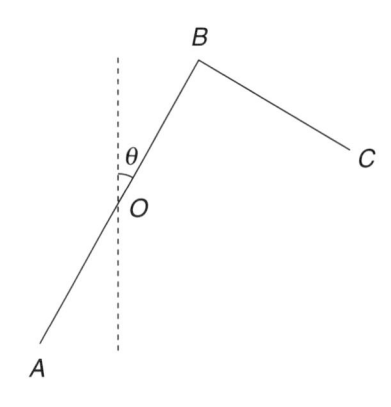

(5)

WJEC

3 A light rod AB, of length $3a$, is smoothly hinged to a vertical wall at A. A load of weight W is suspended from the rod at the point C, which is a distance $2a$ from A. The rod is held at rest in a horizontal position by a light inextensible string attached at B, and to a point D on the wall at a distance $2a$ vertically above A.

(a) Explain what the phrase *light rod* tells you to assume about the weight of the rod. (1)

(b) Show that the tension in the string is $\frac{1}{3}W\sqrt{13}$. (3)

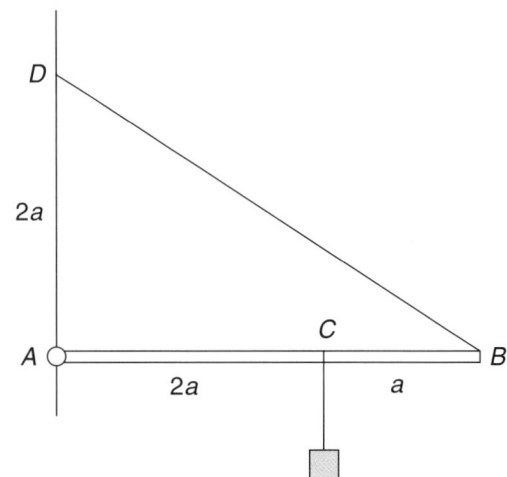

(c) Find, in terms of W, the horizontal and vertical components of the reaction at A. (4)

UODLE

4 A step ladder may be modelled as two equal uniform rods *AB* and *BC* each of length 2 m and of mass 10 kg, freely hinged at *B* and braced by a light, inextensible string attached to the midpoint of each rod. The step ladder is resting on a smooth horizontal floor and the angle *ABC* is 40°.

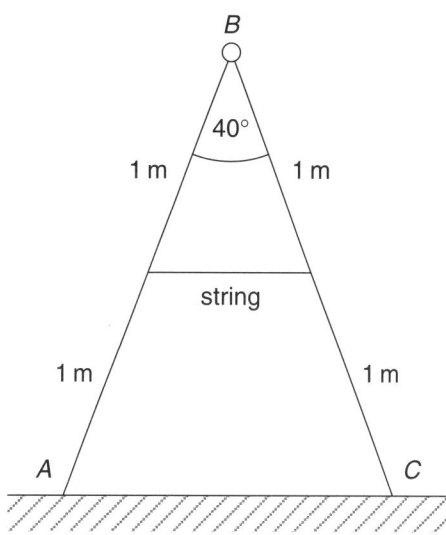

  (i) Explain briefly why the internal forces in the hinge at *B* are horizontal. (2)

  (ii) Draw a diagram showing all four forces acting on the rod *BC*, including those acting in the hinge and the string. (2)

  (iii) By resolving and by taking moments about a suitable point, or otherwise, calculate the tension in the string. (4)

  (iv) Suppose that the floor is rough and the string is cut. What is the least value of the coefficient of friction between the rods and the floor so that the step ladder remains in equilibrium with the same angle *ABC*? (6)

Oxford & Cambridge

5 The diagram shows a framework that is smoothly hinged to a fixed point at *D*. The points *A*, *B*, *C* and *D* form a square, that lies in a vertical plane. A vertical force of 100 N is applied at *A* as shown in the diagram.

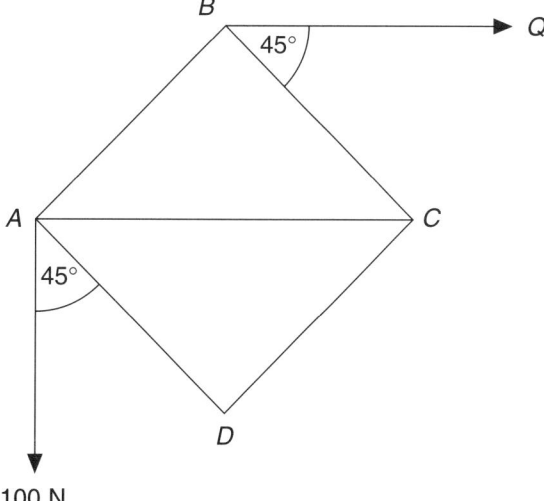

  (a) What two key assumptions need to be made if you are to find the forces in each member of the framework? (2)

  (b) Find the magnitude, *Q*, of the horizontal force that acts at *B*, if the framework is to remain at rest. (2)

  (c) Show that the magnitude of the force in *BC* is $25\sqrt{2}$ N, and find the magnitudes of the forces in the other members of the framework. (7)

  (d) Which members could be replaced by ropes? (1)

AEB

6   The diagram shows a machine component made from a thin uniform circular disc, centre *A* and radius 5 cm, from which a circular hole, centre *B*, has been cut out. The hole has radius 2 cm and *AB* = 2 cm.

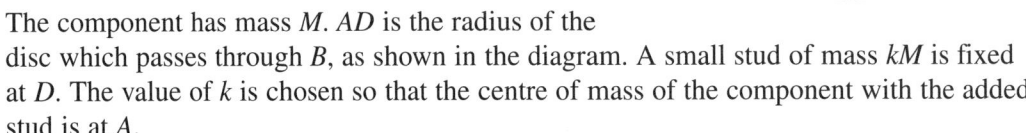

(a) Give a reason why the centre of mass of the component lies on the line through *A* and *B*.

(b) Show that the distance of the centre of mass of the component from *A* is $\frac{8}{21}$ cm.

The component has mass *M*. *AD* is the radius of the disc which passes through *B*, as shown in the diagram. A small stud of mass *kM* is fixed at *D*. The value of *k* is chosen so that the centre of mass of the component with the added stud is at *A*.

(c) Find the value of *k*.   (10)
London Examinations

7

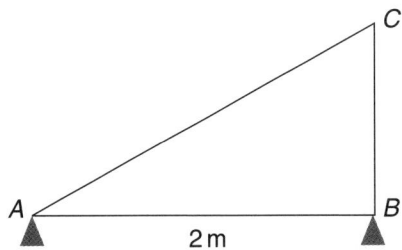

A uniform triangular plate *ABC* has mass 60 kg. The triangle *ABC* is right-angled at *B* and *AB* = 2 m. The plate is vertical and rests on two supports at *A* and *B*, with *AB* horizontal (see diagram).

(i)   State the horizontal distance of the centre of mass of the plate from *A*.   (1)

(ii) Find the magnitudes of the forces on the supports at *A* and *B*.   (4)
UCLES

8   A uniform rectangular metal plate *ABCD*, where *AB* = 6 cm and *BC* = 2 cm, has mass *M*. Four particles of mass *m*, *m*, 3*m* and 3*m* are attached to the points *A*, *B*, *C* and *D* respectively. When the loaded plate is suspended freely from the point *A* and hangs in equilibrium, *AB* makes an angle α with the downward vertical, as shown in the diagram, where $\sin \alpha = \frac{5}{13}$.

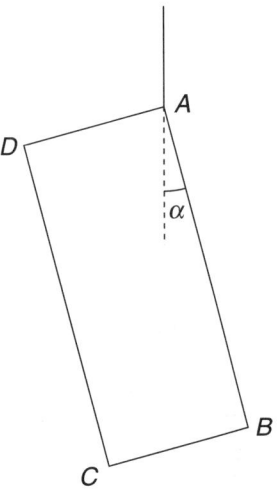

(a) State the distance of the centre of mass of the loaded plate from *AD*.

(b) Show that the distance of the centre of mass of the loaded plate from *AB* is 1.25 cm.

(c) Hence find *m* in terms of *M*.

(13)
London Examinations

9 The diagram shows a uniform solid consisting of a cylinder of radius R and length $2R$, which has P and Q as the centre of its circular ends, and a cone of base R, height $2R$, joined to the end whose centre is Q.

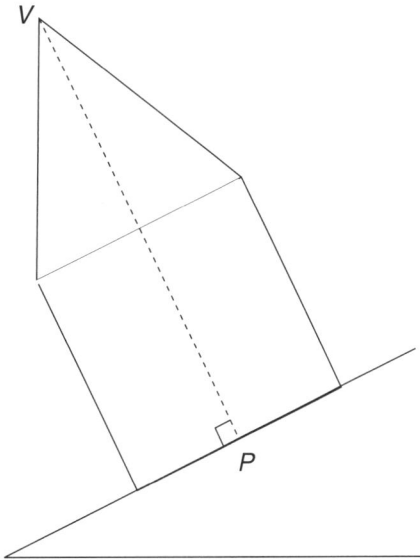

(i) Show that the centre of gravity of the resulting body is at a distance of $\dfrac{11R}{8}$ from P. (9)

The body is now placed, as shown, on a rough plane. The inclination of the plane to the horizontal is gradually increased.

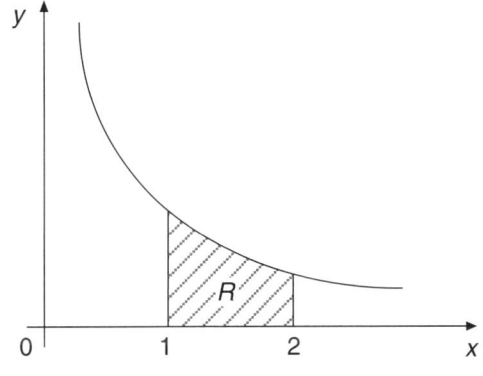

(ii) Find the inclination of the plane when the body is just about to topple. (6)

NICCEA

10 The region R is bounded by the curve with the equation $y = \dfrac{1}{x}$, the lines $x = 1$, $x = 2$ and the x-axis, as shown in the diagram.

The unit of length on both axes is 1 m. A solid plinth is made in the shape of a solid formed by rotating R through 2π about the x-axis.

(a) Show that the volume of the plinth is $\dfrac{\pi}{2}$ m³.

(b) Find the distance of the centre of mass of the plinth from its larger plane face, giving your answer in cm to the nearest cm.

(10)

London Examinations

8 *Circular motion*

A particle moves in a *straight line* with constant velocity unless it is acted on by an external force. Therefore a particle moving on a *circular path* has a force acting on it which has a component along the normal to the path, i.e. along the radius.

Uniform circular motion

When a particle P moves on a circle centre O its position can be determined by measuring the angle θ between the radius OP and a fixed radius OA.

This angle is measured in *radians*. (2π radians $= 360°$)

As P moves around the circle θ changes. The rate of change of θ with respect to time is the *angular speed ω* and this is measured in radians per second, or revolutions per minute.

$$\omega = \frac{d\theta}{dt} = \dot{\theta}.$$

In a uniform circular motion ω is constant, and at time t, $\theta = \omega t$.

If the radius of the circle is r the particle P may be considered to be at the point

$x = r \cos \omega t$, $y = r \sin \omega t$ relative to the axes Ox and Oy, at time t.

$\therefore$ $\mathbf{OP} = r \cos \omega t\, \mathbf{i} + r \sin \omega t\, \mathbf{j}$

Differentiating this gives

$$\mathbf{v} = -r\omega \sin \omega t\, \mathbf{i} + r\omega \cos \omega t\, \mathbf{j}$$

Therefore the magnitude of the linear velocity $\mathbf{v}$ is

$$\sqrt{r^2\omega^2 \sin^2 \omega t + r^2\omega^2 \cos^2 \omega t}\ ,$$

i.e. $|\mathbf{v}| = r\omega$, and its direction is tangential to the circle.

Differentiating further gives

$$\mathbf{a} = -r\omega^2 \cos \omega t\, \mathbf{i} - r\omega^2 \sin \omega t\, \mathbf{j} \quad \therefore \mathbf{a} = -\omega^2\, \mathbf{OP}$$

The magnitude of $\mathbf{a}$ is $|\mathbf{a}| = r\omega^2$ and $\mathbf{a}$ is directed towards the centre of the circle.

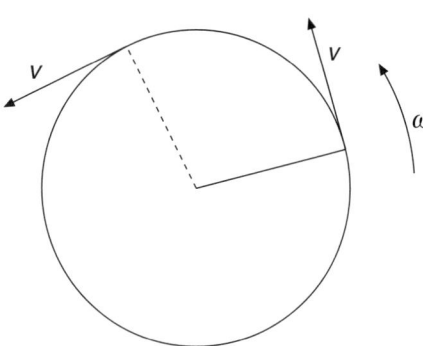

Since $v = r\omega \Rightarrow \omega = \dfrac{v}{r}$ we also obtain $|\mathbf{a}| = \dfrac{v^2}{r}$.

Motion in a horizontal circle

If the particle has mass m, the force necessary to produce this acceleration has magnitude $mr\omega^2$

or $m\,\dfrac{v^2}{r}$, and it is directed towards the centre of the circle. This force may be provided by the

tension T in a string of length l, for example. In this case $T = m\dfrac{v^2}{l}$.

Other problems usually require two equations. One equation is obtained from resolving vertically. The other is obtained from the equation of motion in the horizontal direction. One important example is the conical pendulum.

The conical pendulum

The conical pendulum consists of a particle attached to a fixed point O by a string of length l. The particle moves in a horizontal circle below the level of O and the string generates the surface of a cone. The particle shown in the figure has mass m, the tension in the string is T and the angular speed of the particle is ω. The vertical forces balance and so $T \cos \alpha = mg$.

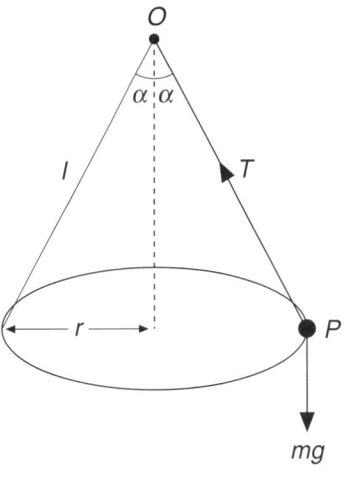

The particle moves in a horizontal circle, so the equation of motion horizontally is $T \sin \alpha = mr\omega^2$.

As $r = l \sin \alpha$ this results in the equation

$$T \sin \alpha = m(l \sin \alpha)\omega^2$$

From these equations $T = ml\omega^2$ and

$$l\omega^2 \cos \alpha = g \implies l\cos \alpha = \frac{g}{\omega^2}.$$

These results should not be quoted but should be obtained from writing down equations for the horizontal and vertical motion.

Other typical examples include the motion of a train around a banked track and of a cyclist turning a corner.

Motion in a vertical circle

The problems encountered usually involve a particle subject to *two* forces, the weight of the particle and a force along the radius of the circular path. As the radial force is at right angles to the direction of motion of the particle it does no work. This means that the sum of the kinetic energy and potential energy of the particle is constant.

Conservation of energy gives a first equation and an equation of motion in the radial direction gives a second equation.

Consider a particle P of mass m attached to one end of a light *rod* of length l. The initial speed of P is u and the tension in the rod when OP makes angle θ with the upward vertical is T.

The equation of motion in the radial direction is

$$mg \cos \theta + T = m\frac{v^2}{l}$$

The conservation of energy equation is

$$\tfrac{1}{2} mu^2 - \tfrac{1}{2} mv^2 = mgl(1 + \cos \theta)$$

The condition for P to describe complete circles is that $v \geq 0$ when $\theta = 0$. Thus $u^2 \geq 4gl$ for complete circles. The tension in a *rod* can be positive, zero, or negative. In the latter case it is a thrust, rather than a tension, but the value of the tension places no constraints on the motion.

If however the particle is attached to a light inelastic *string* the least value of the tension is zero, and if the tension becomes zero before the particle reaches the top of the circle, then the string will become slack and the particle will cease to move on a vertical circle. It will instead move as a projectile under gravity, until the string again becomes taut. The condition for complete circles, in the case of a particle attached to a string, is that $T \geq 0$ at $\theta = 0$.

The condition for complete circles is that $u^2 \geq 5gl$.

Further examples include particles moving in contact with a spherical surface. In such cases the normal reaction force replaces the previous tension force and when the reaction is zero the particle leaves the surface. If, however, the problem is that of a bead threaded on a circular wire, the bead cannot leave the wire and the reaction can act in the other direction and become negative.

1 A small coin is placed flat on a record turntable. The coin is at a distance of 10 cm from the axis of the turntable (see diagram). The turntable is rotating at a constant speed of 33 revolutions per minute, and the coin is not moving on the turntable.

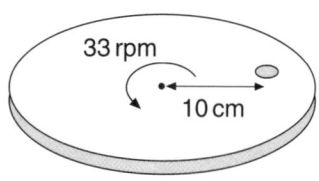

(i) Show that the angular speed of the turntable is 3.46 rad s^{-1}, correct to three significant figures. (1)

(ii) Find, in m s^{-1}, the speed of the coin. (2)

(iii) Find the acceleration of the coin, giving its magnitude in m s^{-2} and stating its direction. (3)

UCLES

2 The car, shown in the diagram, is travelling round a circular bend on a road banked at an angle α to the horizontal. The car may be modelled as a particle moving in a horizontal circle of radius 120 metres.

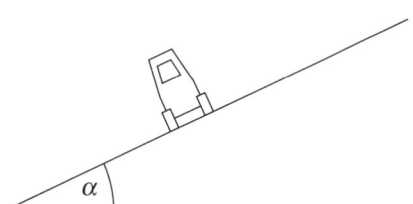

When the car is moving at a constant speed of 20 m s^{-1} there is no frictional force up or down the slope.

Find the angle α, giving your answer in degrees correct to one decimal place. (4)

UODLE

3 A vertical post is fixed in the ground. A tennis ball, of mass 0.05 kg, is attached to the top of the post by a string. To control the height of the ball a second string, of the same length as the first, joins the ball to the post at ground level. The ball is moving, with constant speed of 5 m s^{-1}, in a horizontal circle of radius 1.2 m. Each string is taut and inclined at a constant angle of 55° to the vertical, as shown in the diagram. The modelling assumptions made are that both strings are light and inextensible, and that there is no

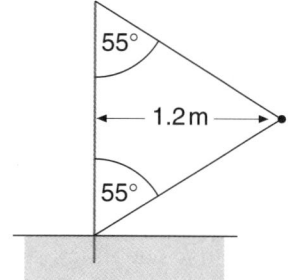

air resistance. Find the tensions in the strings, giving your answers in newtons, correct to two decimal places. (7)

UCLES

4

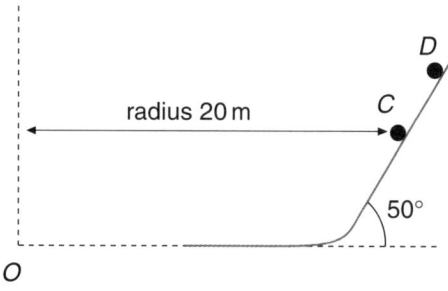

The diagram shows the cross-section of a banked circular cycle racing track, in a vertical plane through its centre *O*. A cyclist *C* moves at a constant speed on the steepest part of the track in a horizontal circular path of radius 20 m. The mass of the cyclist and his machine is 80 kg and the angle between the steepest part of the track and the horizontal is 50°. By using a model in which the cyclist and his machine are considered as a single particle and in which the track is smooth:

(i) calculate the normal component of the contact force on the machine due to the track,

(continued)

(ii) show that the magnitude of the acceleration of C is 11.7 m s^{-2}, correct to 3 significant figures,

(iii) calculate the speed of C. (8)

Another cyclist D moves at constant speed on the steepest part of the track in a higher horizontal circular path. Determine which of the cyclists C and D completes one circuit of the track in the shorter time. (4)

UCLES

5 A particle of mass m is hanging at rest from a fixed point O by an inelastic string of length a. The particle is given a horizontal impulse which causes it to move with a speed U.

(i) Determine the value of U which will cause the particle to just reach the horizontal level of O. (4)

(ii) Determine the smallest value of U which will cause the particle to perform complete circular revolutions about O. (6)

In the case when $U = \sqrt{\dfrac{7}{2}ag}$,

(iii) locate the point at which the particle ceases to follow a circular path, (8)

(iv) determine its speed at this point. (2)

Oxford & Cambridge

6 A marble P of mass m moves in a vertical circle, with centre O and radius a, inside a hollow cylinder. The axis of the cylinder is horizontal and perpendicular to the plane of the motion of P.

In a model of the situation, it is assumed that the marble is a particle, the cylinder is smooth, and the only forces acting on the marble are its weight and the force exerted by the cylinder.

Initially P is at the point A on the circle, where OA is horizontal, and is moving with a vertical speed of $\sqrt{5ga}$ in a downwards direction. In the subsequent motion, when OP makes an angle θ with OA, the speed of P is v and the force exerted by the cylinder on P has magnitude R.

(a) Find an expression for v^2 in terms of a, g and θ.

(b) Find an expression for R in terms of m, g and θ.

(c) Show that P will make complete revolutions without losing contact with the surface of the cylinder.

(d) Find the minimum value of the speed of P.

(e) State two forces which have been ignored in the above model. (14)

London Examinations

9 *Simple Harmonic Motion*

If you need to revise this subject more thoroughly, see the relevant topics in the *Letts* **A-level Mathematics Study Guide.**

In Simple Harmonic Motion (S.H.M.) a particle oscillates in a straight line about a fixed point O, called the **centre of oscillation**. The acceleration of the particle is always directed towards O, and the magnitude of the acceleration is proportional to the distance, x, of the particle from O.

The acceleration of the particle is $\ddot{x}$, and $\ddot{x} = -\omega^2 x$, where ω is constant.

This equation may be rewritten as $v \dfrac{\mathrm{d}v}{\mathrm{d}x} = -\omega^2 x$.

The solution of this equation gives $v^2 = \omega^2(a^2 - x^2)$, where a is the amplitude of the S.H.M., i.e. a is the maximum displacement of the particle from O.

The particle has maximum speed at O, when $v = \pm a\omega$.

Since $v = \dfrac{\mathrm{d}x}{\mathrm{d}t} = \pm \omega \sqrt{a^2 - x^2}$, x can be expressed in terms of t.

The solution of this differential equation gives $x = a \sin(\omega t + \varepsilon)$ which is the general solution of the S.H.M.

The particular solutions are $x = a \sin \omega t$ if $x = 0$ when $t = 0$, and $x = a \cos \omega t$ if $x = a$ when $t = 0$.

The general solution is used for other initial conditions.

The time taken to complete an oscillation is called the **period** of the motion (T), and this is $\dfrac{2\pi}{\omega}$.

Projection of circular motion

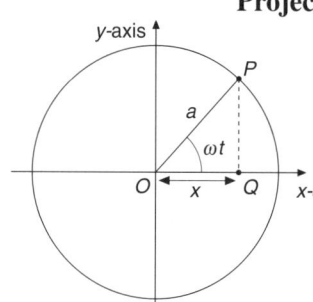

A particle P moves around a circle of radius a with uniform angular speed ω.

A second particle Q is the foot of the perpendicular from P to the x-axis.

The acceleration of P has magnitude $r\omega^2$ and is directed along PO.

The acceleration of Q is the resolved part of this acceleration in the x direction,

i.e. $-r\omega^2 \cos \omega t = -r\omega^2 \dfrac{x}{r} = -\omega^2 x$

The motion of Q is S.H.M. with period $\dfrac{2\pi}{\omega}$ and amplitude a.

The circle associated with a particular S.H.M. is the **reference circle**. It can be used to calculate the time taken to move between two specific points in an oscillation. There are many examples of S.H.M. and some are included in Unit 10, as they relate to elastic strings and springs.

The simple pendulum

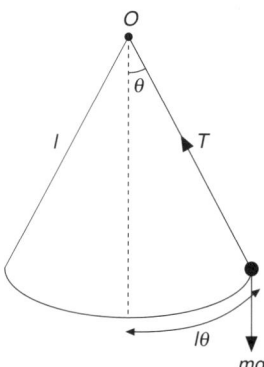

A simple pendulum consists of a small bob hanging from a long thread attached to a fixed point. The bob and string are modelled by a particle and a light inextensible string. Consider a particle of mass m connected to one end of a light inelastic string of length l whose other end is attached to a fixed point O. The particle oscillates in a circular arc. It is assumed that the oscillations are small, and air resistance is neglected. The length of the arc, measured from the lowest point, is given by the equation $s = l\theta$,

and so the acceleration along the tangent is $\dfrac{\mathrm{d}^2 s}{\mathrm{d}t^2} = l \dfrac{\mathrm{d}^2 \theta}{\mathrm{d}t^2}$.

The equation of motion along the tangent is $ml \dfrac{\mathrm{d}^2 \theta}{\mathrm{d}t^2} = -mg \sin \theta$.

If θ is small then $\sin \theta \approx \theta$ and therefore $\dfrac{\mathrm{d}^2 \theta}{\mathrm{d}t^2} = -\dfrac{g}{l} \theta$.

This is Simple Harmonic Motion with period $2\pi \sqrt{\dfrac{l}{g}}$.

1 A machine component consists of a small piston of mass 0.2 kg moving inside a fixed horizontal cylinder. The piston, which is modelled as a particle, moves with simple harmonic motion and the total distance from one end of each oscillation to the other is 0.05 m. Given that the speed of the piston must not exceed $20\,\text{m s}^{-1}$, find

(a) the maximum number of oscillations per second which the piston can perform,

(b) the maximum horizontal force that can be experienced by the piston. (8)

London Examinations

2 The positions of the points B, B', A, A' and O are shown on the diagram below.

$$\begin{array}{ccccccc} B' & A' & O & A & B & \\ \vdash & \vdash & \vdash & \vdash & \vdash & \dashrightarrow x \\ -b & -a & & a & b & \end{array}$$

A particle is released from rest at the point B at time $t = 0$ and moves in a Simple Harmonic Motion centre O with period $\dfrac{2\pi}{\omega}$.

(i) Write down an expression for the displacement x from O of the particle at time t. (2)

(ii) Write down an expression for the speed v of the particle as it passes the point A and

show that $b = \sqrt{a^2 + \dfrac{v^2}{\omega^2}}$. (2)

(iii) Show that the time taken by the particle to travel directly from A to A' is given by

$$\frac{1}{\omega}\left[\pi - 2\arccos\left(\frac{a}{b}\right)\right].$$ (3)

A second particle is released from rest at the point A at time $t = 0$. It moves in a Simple Harmonic Motion centre O with period $\dfrac{2\pi}{\omega}$.

(iv) How much longer does this particle take to travel from A to A' than the first one? (1)

A machine component is subject to a variable force which causes it to move with Simple Harmonic Motion. When released from rest at point A it travels 0.4 m before coming momentarily to rest after 0.1 seconds. It is desired to shorten the time of the motion of the component over the 0.4 m and this is to be achieved by giving the component an initial speed of $2\,\text{m s}^{-1}$.

(v) Assuming that the same law of force applies to the new motion, calculate the time saved. [*Hint*: use the result obtained in (ii) to find the amplitude of the Simple Harmonic Motion the component would have if it were released from rest and passed through point A at $2\,\text{m s}^{-1}$.] (4)

Oxford & Cambridge

3 A particle of mass 0.2 kg is suspended from a fixed point A by a light inextensible string of length 25 cm. The particle is set in motion in a vertical plane through A so that it performs small oscillations with the string taut.

(a) (i) Show that the equation of motion of the particle can be expressed in the form

$$\ddot{\theta} \approx -\omega^2\theta, \text{ where } \theta \text{ is the angle the string makes with the vertical through } A. \quad (3)$$

(ii) State the numerical value of ω^2. (1)

(b) Find the period of the small oscillations, giving your answer correct to three significant figures. (2)

UODLE

10 *Elastic strings and springs*

Hooke's Law states that **the tension in a stretched string or spring is proportional to the extension**, provided that the elastic limit is not reached, i.e. $T = kx$ where k is the stiffness of the string or spring. This equation is often rewritten as

$T = \dfrac{\lambda x}{l}$ where T is the tension,

$\qquad\qquad\qquad$ x is the extension

$\qquad\qquad\qquad$ l is the natural length of the string or spring.

The constant λ is called the **modulus of elasticity** and is measured in newtons. (It is the force required to double the length of the unstretched string or spring.)

Hooke's Law also applies to springs under compression, in which case the thrust in the spring is proportional to the compression.

The elastic potential energy

The work done in stretching a string or spring from a length $(l + x_1)$ to $(l + x_2)$ is $\displaystyle\int_{x_1}^{x_2} T \, dx$.

As $T = \dfrac{\lambda x}{l}$, the work done $W = \displaystyle\int_{x_1}^{x_2} \dfrac{\lambda x}{l} \, dx = \left[\dfrac{\lambda x^2}{2l} \right]_{x_1}^{x_2} = \dfrac{1}{2} \dfrac{\lambda}{l} \left(x_2{}^2 - x_1{}^2 \right).$

The energy stored in a stretched string is $\dfrac{\lambda x^2}{2l}$ or $\dfrac{1}{2} kx^2$. This is called its **elastic potential energy**.

The work–energy principle

The total change of the energy of a system is equal to the work done by any external forces acting on the system. If the only external force acting is the weight of the particle connected to the string or spring then conservation of energy can be used. The total energy, made up of the kinetic energy, gravitational potential energy and elastic potential energy, remains constant. Equating the total initial energy with the total energy at some specified time can often simplify calculations, particularly if one or more of these energies is zero.

Simple Harmonic Motion

The motion of particles attached to elastic strings or springs can give rise to Simple Harmonic Motion as shown in the following examples.

If you need to revise this subject more thoroughly, see the relevant topics in the *Letts* A-level Mathematics Study Guide.

- *Motion of a stretched horizontal spring*

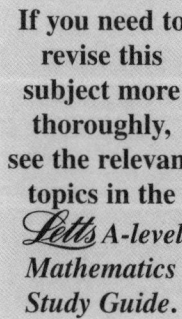

The motion is S.H.M. with $m\dfrac{d^2 x}{dt^2} = m\ddot{x} = \dfrac{-\lambda x}{l}$ and the period of oscillation $= 2\pi \sqrt{\dfrac{ml}{\lambda}}$.

- *Motion of a vertical spring*

In equilibrium the extension is e. $T_{equilib} = mg$ and $T_{equilib} = \dfrac{\lambda e}{l}$, $\therefore mg = \dfrac{\lambda e}{l}$.

Under a further extension x : $\qquad T_{new} = \dfrac{\lambda (e+x)}{l}$ $\therefore m\ddot{x} = mg - T_{new} = \dfrac{-\lambda x}{l}$.

$\therefore$ period of oscillation $= 2\pi \sqrt{\dfrac{ml}{\lambda}}$

- *Motion of an elastic string*

For an elastic string the approach is similar. However the S.H.M. will cease when the string is unstretched.

1 A particle *P* of mass 0.13 kg moves on a smooth horizontal table. The particle is attached to one end of a light elastic string with natural length 1.5 m and modulus of elasticity 78 N. The other end of the string is attached to a fixed point *O* of the table. The particle is released from rest at a distance 2.2 m from *O*. Ignoring air resistance, calculate the speed of *P* when the string becomes slack. (3)

UCLES

2 A particle *P* of mass 0.2 kg is suspended in equilibrium from a fixed point *O* by a <u>light</u> <u>extensible string of natural length</u> 0.4 m. State which one of the words underlined enables you to assume that the tension is the same at all points of the string. (1)

Given that the modulus of elasticity of the string is 10 N, find the distance *OP*. (3)

WJEC

3 A particle of mass 0.5 kg is suspended from a fixed point *O* by a light elastic string of natural length 1.5 metres and stiffness 8 N m^{-1}. The particle is released from rest at *O* and next comes to rest at the point *A*.
(a) When the particle is at the point *A*, the extension in the string is *x* metres. Show, using energy considerations, that *x* satisfies the equation $8x^2 - 10x - 15 = 0$. (4)
(b) Hence find the distance *OA*, giving your answer in metres, correct to three significant figures. (3)

UODLE

4 A light elastic string of modulus 32 N and natural length 0.8 m has one end attached to a fixed point *O*. A particle of mass 0.5 kg is attached to the other end. The particle is released from rest at *O*.
(i) Show that, when the extension in the string is *x* metres, the elastic potential energy in the string is $20x^2$ joules. (2)
(ii) Calculate the value of *x* when the particle is at its lowest point. (4)
(iii) Calculate the speed of the particle when it is at the point 0.9 m directly below *O*. (5)

UCLES

5 A small catapult consists of a light elastic string fastened to two fixed points *A* and *B*, with the line *AB* horizontal and the distance *AB* = 30 cm. The natural length of the string is 20 cm. When the string forms a straight line between *A* and *B*, the tension in the string is 150 N.
(a) Find, in N, the modulus of elasticity of the string.

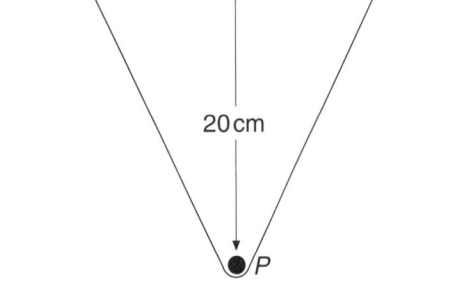

A small light leather pouch *P* is fixed to the midpoint of the string. The pouch is now pulled back horizontally a distance of 20 cm in a direction perpendicular to *AB*, so that *A*, *B* and *P* all lie in the same horizontal plane. The diagram shows a view of the catapult *from above*.
(b) Find the magnitude of the horizontal force required to hold the pouch in equilibrium in this position.

A small stone of mass 0.1 kg is placed in the pouch and held in the position shown in the diagram. The pouch is then released from this position. By considering the energy of the system,
(c) find, in m s^{-1}, the horizontal speed which the stone has when it crosses the line *AB*. (Any vertical motion of the stone can be assumed to be so small that it may be neglected.) (15)

London Examinations

QUESTIONS

6   A particle P of mass m is attached to one end of a light elastic string with natural length L and modulus of elasticity $4mg$. The other end of the string is attached to a fixed point A. The particle describes a horizontal circle with constant speed v. The centre of this circle is the point O, where O is vertically below A and the distance $OA = L$, as shown in the diagram.

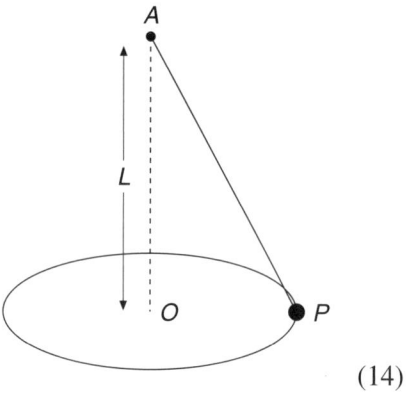

(a) Show that the extension of the string is $\dfrac{L}{3}$.

(b) Show that $v = \frac{1}{3}\sqrt{7gL}$.

(14)

London Examinations

7   A bungee jumper of mass m is attached to a length of elastic rope. The bungee jumper steps off a high platform to which the other end of the elastic rope is attached. After leaving the platform, the bungee jumper first comes to rest momentarily when the elastic rope has doubled its length.

Assume that:

(I)   The bungee jumper can be modelled as a particle and that there is no air resistance present.

(II) The tension in the rope is given by $T = \dfrac{\lambda x}{l}$, where l is the unstretched or natural length of the rope, x the extension of the rope and λ a constant that depends on the characteristics of the rope.

(a) Use integration to show that the work done in stretching the rope to twice its natural length is $\dfrac{\lambda l}{2}$, and hence show that $\lambda = 4mg$. (4)

(b) Prove that the maximum speed of the bungee jumper is $\dfrac{3}{2}\sqrt{gl}$. (5)

(c) In reality, the bungee jumper eventually comes to rest and remains at a point A. State the distance between A and the platform that the bungee jumper stepped off. (1)

(d) Shortly before coming to rest at A the bungee jumper described vertical oscillations of small amplitude about A, during which the rope remained taut.

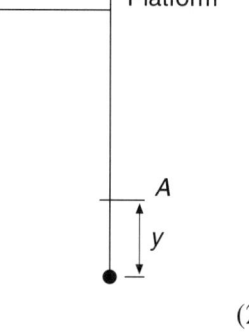

(i)   Find an expression for the acceleration of the bungee jumper during this motion in terms of y, his displacement from A, assuming that the rope remains taut. (3)

(ii) Hence state the period of these oscillations. (1)

(e) Use your answer to (d) (ii) to state the factor on which the period of the oscillations of the bungee jumper depends. Hence state what happens to the period of his oscillations as their amplitude slowly decreases. (2)

AEB

Answers

1 VECTORS

Answer	Mark	Examiner's tip

1 Use resolved parts of forces
$R\cos\theta = 5 + 7\cos 55°$ and | **1** | Use diagram on question paper.
| | Compare forces in two
$R\sin\theta = 7\sin 55°$ | **1** | perpendicular directions.

Square and add the equations to give R^2 and square root both sides. | **1** |

$\underline{R = 11\text{ N}}$ | **1** | 2 s.f. is required for this board.

Divide the equations to give $\tan\theta$. | **1** | Always use unrounded figures in

$\underline{\theta = 32°}$ | **1** | calculations; to nearest degree.

2

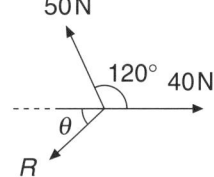

50 N
120° 40 N
θ
R

Draw a clear labelled diagram showing the forces. Try to put R in roughly the right direction and mark θ on your diagram.

$40 - 50\cos 60° - R\cos\theta = 0$ | **2** | System is in equilibrium so net effect in any direction is zero.

$R\cos\theta = 15$ | **1** | Simplify your equation.

$50\sin 60° - R\sin\theta = 0$ | **2** | Consider a perpendicular direction.

$R\sin\theta = 25\sqrt{3}$ | **1** | Again simplify your equation.

Square and add the equations to give R^2, and square root: | **1** |

$R = \sqrt{15^2 + (25\sqrt{3})^2} = \sqrt{2100} = \underline{45.8\text{ N}}$ | **1** |

Divide the equations to give $\tan\theta = \dfrac{5\sqrt{3}}{3}$ | **2** |

$\underline{\theta = 70.9°}$ | **1** | As above.

3 (a) (i) $|\mathbf{a}| = \sqrt{2^2 + 1^2 + q^2}$ | **1** | Using Pythagoras in 3D.

$|\mathbf{b}| = \sqrt{q^2 + (-2)^2 + (2q)^2}$ | **1** |

$5 + q^2 = 4 + 5q^2$ | | Equate magnitudes, square both sides and simplify.

$\underline{q = \pm\tfrac{1}{2}}$ | **2** | q can be positive or negative.

(ii) $\underline{\mathbf{R} = \mathbf{a} + \mathbf{b} = 5\mathbf{i} - \mathbf{j} + 9\mathbf{k}}$ | **2** | Add the components to find the resultant.

(iii) $|\mathbf{a}| = \sqrt{2^2 + 1^2 + 2^2} = 3$ | **1** |

$\underline{\hat{\mathbf{a}} = \mathbf{a} \div 3 = \tfrac{2}{3}\mathbf{i} + \tfrac{1}{3}\mathbf{j} + \tfrac{2}{3}\mathbf{k}}$ | **2** | Unit vector = vector ÷ its magnitude

Answer	Mark	Examiner's tip

(b) Resultant $= \overrightarrow{AC} + \overrightarrow{CB} + \overrightarrow{AB}$ **2**

$\qquad\qquad = \overrightarrow{AB} + \overrightarrow{AB} = 2\,\overrightarrow{AB}$

i.e. <u>$2F$ parallel to AB</u>   **2**

Using $\overrightarrow{AC} + \overrightarrow{CB} = \overrightarrow{AB}$

4 (a) $(120\cos 60° + 100 + 90\cos 70°)\mathbf{i} +$

$(120\sin 60° - 90\sin 70°)\mathbf{j}$

<u>$= 191\mathbf{i} + 19.4\mathbf{j}$</u>   **2**

Resolve each force into $\mathbf{i}$ and $\mathbf{j}$ components; add the components to find the resultant

(b) $\mathbf{F} + 191\mathbf{i} + 19.4\mathbf{j} = 0$

$\qquad\qquad \mathbf{F} = - (191\mathbf{i} + 19.4\mathbf{j})$

$\qquad\qquad\quad$ <u>$= -191\mathbf{i} - 19.4\mathbf{j}$</u>   **1**

Pole is now in equilibrium.

(c) $|\mathbf{F}| = \sqrt{(-191)^2 + (-19.4)^2}$ **1**

$\qquad\quad$ <u>$= 192\,\text{N}$</u>   **1**

By Pythagoras.

To 3 s.f.

(d)   **2**

191

19.4

$\mathbf{i}$

(e)

200

θ

192

Draw a diagram.

$200\cos\theta = 192$ **1**

<u>$\theta = 16.3°$</u>   **1**

The horizontal component must equal 192 in order to maintain the pole in equilibrium.

5 (a) Line has direction $\begin{pmatrix} 6 \\ -2 \\ 6 \end{pmatrix}$.

Note that this exam board uses column vectors.

Line joining $(-1, 4, 5)$ to $(1, 3, 7)$ has

direction $\begin{pmatrix} 2 \\ -1 \\ 2 \end{pmatrix}$.

Since $\begin{pmatrix} 2 \\ -1 \\ 2 \end{pmatrix}$ is not a multiple of $\begin{pmatrix} 6 \\ -2 \\ 6 \end{pmatrix}$,   **1**

points are <u>not</u> in a straight line.   **1**

Direction vectors would be parallel if points were collinear.

(b) $(2 \times -3) + (3 \times 0) + (4 \times 1) = \sqrt{29} \times \sqrt{10}\cos\theta$ **1**

<u>$\theta = 1.69\,\text{rad}$</u>   **1**

Using the scalar product calculated in two different ways.

(c) $\mathbf{q} - \mathbf{p} = 2\mathbf{i}$ and $\mathbf{p} - 2\mathbf{r} = 3\mathbf{i}$

$\qquad 3(\mathbf{q} - \mathbf{p}) = 2(\mathbf{p} - 2\mathbf{r})$ **1**

<u>$5\mathbf{p} - 3\mathbf{q} - 4\mathbf{r} = 0$</u>   **1**

Two equations in $\mathbf{i}$ (or $\mathbf{j}$).

Eliminate $\mathbf{i}$ (or $\mathbf{j}$).

Simplify.

6 (a) $800\sin 10° - 500\sin\alpha° = 0$ **2**

<u>$\alpha = 16$</u>   **2**

Resolving perpendicular to the direction of motion.

Simplify and solve, to nearest whole number.

Answer	Mark	Examiner's tip
(b) $800\cos 10° + 500\cos \alpha° - F = 0$	2	Resolving in direction of motion. Note that constant speed $\Rightarrow a = 0$
$\underline{F = 1270\ \text{N}}$	1	Solve to 3 s.f.
(c) Rock is small or speed slow.	1	
(d) Ropes inclined at angle to horizontal.	1	

7 (a) 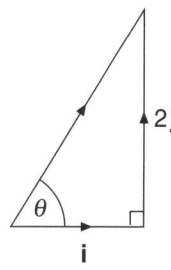 $\tan\theta = 2 \Rightarrow \underline{\theta = 63°}$

Draw a diagram, marking the angle required.

To nearest degree. **2**

(b) $\mathbf{R} = \mathbf{F}_1 + \mathbf{F}_2 = (2 + \lambda)\mathbf{i} + (3 + \mu)\mathbf{j}$	1	Add the components.
$2(2 + \lambda) = (3 + \mu)$	1	Since $\mathbf{R}$ is parallel to $\mathbf{i} + 2\mathbf{j}$.
$\underline{2\lambda - \mu + 1 = 0}$	1	Simplifying to required answer.
(c) $\mathbf{F}_2$ parallel to $\mathbf{j} \Rightarrow \lambda = 0$	1	Component of $\mathbf{i}$ must be zero.
$\lambda = 0 \Rightarrow \mu = 1$	1	Using result from (b).
$\mathbf{R} = 2\mathbf{i} + 4\mathbf{j}\ ;\ |\mathbf{R}| = \sqrt{4 + 16}$	1	By Pythagoras.
$\underline{= 4.47\ \text{N}}$	1	To 3 s.f.
8 (a) $\underline{\mathbf{r}_D = t(3\mathbf{i} + 5\mathbf{j})}$	1	David starts at the origin.
$\underline{\mathbf{r}_C = (10\mathbf{i} + 8\mathbf{j}) + t(3\mathbf{i} - 4\mathbf{j})}$	2	Colin starts at point with position vector $(10\mathbf{i} + 8\mathbf{j})$.
(b) When $t = 2$, $\mathbf{r}_C = (10\mathbf{i} + 8\mathbf{j}) + 2(3\mathbf{i} - 4\mathbf{j})$	1	Using answer to (a).
$= 16\mathbf{i}$	1	
$\mathbf{r}_B = 2 \times 8\mathbf{i} = 16\mathbf{i}$	1	Ball starts at O.
Hence $\mathbf{r}_B = \mathbf{r}_C$ when $t = 2$.	1	Thus Colin intercepts ball at $(16, 0)$.
(c) When $t = 4$, $\mathbf{r}_D = 4(3\mathbf{i} + 5\mathbf{j}) = 12\mathbf{i} + 20\mathbf{j}$	2	Using answer to (a).
After 2 s, $\mathbf{r}_B = 16\mathbf{i} + 2(\lambda\mathbf{i} + \mu\mathbf{j})$		We need position of ball 2 s after Colin has kicked it from $(16, 0)$.
$= (16 + 2\lambda)\mathbf{i} + 2\mu\mathbf{j}$	2	
so, $12 = (16 + 2\lambda) \Rightarrow \underline{\lambda = -2}$	2	Equating coefficients of $\mathbf{i}$ and $\mathbf{j}$.
$20 = 2\mu \Rightarrow \underline{\mu = 10}$	1	
9 (a) At B, $4 = 4\mu \Rightarrow \mu = 1$	1	B is point where BE and BC meet so we can equate $\mathbf{j}$ components.
B is the point $\underline{(7, 4, 7)}$.	1	Putting $\mu = 1$ in equation of BC.
(b) $\underline{\mathbf{r} = \begin{pmatrix} 10 \\ 0 \\ 5 \end{pmatrix} + s\begin{pmatrix} 1 \\ 0 \\ 0 \end{pmatrix}}$	1	Since CF is parallel to BE it has the same direction vector.

Answer	Mark	Examiner's tip
(c) (i) $\overrightarrow{CA} = \begin{pmatrix} 0 \\ 8 \\ 0 \end{pmatrix}$	1	This is a direction vector for the line CA.
$\mathbf{r} = \begin{pmatrix} 10 \\ 0 \\ 5 \end{pmatrix} + t \begin{pmatrix} 0 \\ 8 \\ 0 \end{pmatrix}$	1	Note that this answer is not unique.
(ii) $\overrightarrow{CB} = \begin{pmatrix} -3 \\ 4 \\ 2 \end{pmatrix}$	1	We need angle between $\overrightarrow{CB}$ and $\overrightarrow{CA}$.
$(0 \times -3) + (8 \times 4) + (0 \times 2) = 8 \times \sqrt{29} \cos\theta$	1	Using the scalar product calcu-
$\underline{\theta = 42.0°}$	1	lated in two different ways.

2 KINEMATICS OF A PARTICLE

Answer	Mark	Examiner's tip
1 (a) $\mathbf{F} = 0.1(20\mathbf{i} - 30\mathbf{j})$	2	$\mathbf{F} = m\mathbf{a}$ when $t = 2$.
$= 2\mathbf{i} - 3\mathbf{j}$	1	
$\lvert \mathbf{F} \rvert = \sqrt{2^2 + (-3)^2} = 3.61$ N	2	Using Pythagoras; to 2 d.p.
(b) $\mathbf{v} = \dfrac{5t^2}{2}\mathbf{i} - 10t^{1.5}\mathbf{j} + \mathbf{C}$	3	Integrating $\mathbf{a}$ w.r.t. time.
$\mathbf{C} = 10\mathbf{i}$	1	When $t = 0$, $\mathbf{v} = 10\mathbf{i}$.
(c) When $t = 4$, $v = 50\mathbf{i} - 80\mathbf{j}$	2	
		Draw a simple diagram showing θ clearly marked.
$\tan\theta = 80 \div 50 = 1.6 \Rightarrow \theta = 58°$	2	
$\underline{\text{Bearing is } 148°}$	1	Bearing is from N clockwise.
2 Consider horizontal motion:		Need to resolve the initial velocity.
$120 = 40\cos\theta \times 5$	2	$s = ut$ since $a = 0$ horizontally.
$\theta = 53°$	1	To nearest degree for this board.
Consider vertical motion:		Need to resolve the initial velocity.
$h = 40\sin\theta \times 5 - \frac{1}{2}g \times 25$	2	Use $s = ut + \frac{1}{2}at^2$; use $g = 9.81 \text{ m s}^{-2}$.
$h = 37$	1	To 2 s.f.
Air resistance	1	

Answer	Mark	Examiner's tip

3 (a) (i) $1.8 = \frac{1}{2} \times 10 \times t^2$

  $t = 0.6$ s

  1

  1

Using $s = ut + \frac{1}{2}at^2$ with $u = 0$.

  (ii) $v = 10 \times 0.6 = 6\,\text{ms}^{-1}$

  2

Using $v = u + at$ with $u = 0$.

  (b) (i) $-3 = 3 - 10t \Rightarrow t = 0.6$ s

  2

Using $v = u + at$ with upwards as +*ve* direction.

  (ii) $0^2 = 3^2 + 2 \times -10 \times s$

  $20s = 9 \Rightarrow s = 0.45\,\text{m}$

  1

Using $v^2 = u^2 + 2as$ with upwards as +*ve* direction.

  (c) Air resistance

  1

  (d) (a) (i) Greater

  1

Smaller $g \Rightarrow$ longer to fall.

  (a) (ii) Less

  1

Smaller $g \Rightarrow$ smaller final speed.

  (b) (i) Greater

  1

Smaller $g \Rightarrow$ longer to decelerate and accelerate.

  (b) (ii) Same

  2

Energy of ball reduced by factor of 2^2 by impact with ground (as K.E. is proportional to v^2) and so ball will rebound to quarter of height from which it was released, i.e. $1.8 \div 4 = 0.45$ m.

4 (i) $\mathbf{a} = \frac{1}{15}\,[(10.5 - 0)\mathbf{i} + (-0.9 - 0.6)\mathbf{j}]$

  2

$\mathbf{a} = \dfrac{\mathbf{v} - \mathbf{u}}{t}$ as $\mathbf{a}$ is constant.

  $= (0.7\mathbf{i} - 0.1\mathbf{j})\,\text{ms}^{-2}$

  1

  (ii) $0.6t\,\mathbf{j} + \dfrac{t^2}{2}(0.7\mathbf{i} - 0.1\mathbf{j})$

  2

Using $\mathbf{s} = \mathbf{u}t + \frac{1}{2}\mathbf{a}t^2$.

  $= 0.7\dfrac{t^2}{2}\mathbf{i} + (0.6t - 0.1\dfrac{t^2}{2})\,\mathbf{j}$

Collecting components.

  $0.7\dfrac{t^2}{2} = 0.6t - 0.1\dfrac{t^2}{2}$

  1

To be NE of O, $\mathbf{i}$-component must be equal to $\mathbf{j}$-component.

  $t = 1.5$ s

  1

Solving quadratic.

  (iii) $0.6\mathbf{j} + t(0.7\mathbf{i} - 0.1\mathbf{j})$

  2

Using $\mathbf{v} = \mathbf{u} + \mathbf{a}t$.

  $= 0.7t\,\mathbf{i} + (0.6 - 0.1t)\mathbf{j}$

Collecting components as before.

  $0.7t = 0.6 - 0.1t$

  1

To be travelling NE, $\mathbf{i}$-component must be equal to $\mathbf{j}$-component.

  $t = 0.75$ s

  1

  (iv) When $t = 17$, position is $101.15\mathbf{i} - 4.25\mathbf{j}$

  1

Using formula in part (ii).

  Distance from $O = \sqrt{101.15^2 + (-4.25)^2}$

  1

Using Pythagoras.

  $101.24 > 100$

  1

Verifying that contact lost.

5 (a)

  1

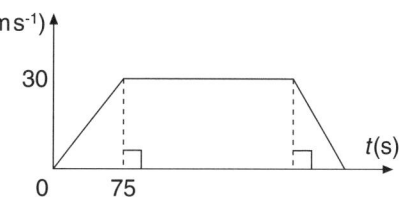

(*continued*)

Answer	Mark	Examiner's tip

(b) a = gradient of graph = $\dfrac{30}{75}$ = $\underline{0.4\,\text{m s}^{-2}}$ **1**

s = area under graph

$= \dfrac{30 \times 75}{2}$ = $\underline{1125\,\text{m}}$ **2**

Area of triangle is required.

(c) Time at constant speed $= \dfrac{22500 - 2250}{30}$ **1**

$= 675\,\text{s}$

Total time $= 675 + 150 = \underline{825\,\text{s}}$ **1**

Need to convert 22.5 km into m; distance covered whilst decelerating is also 1125 m.

(d)

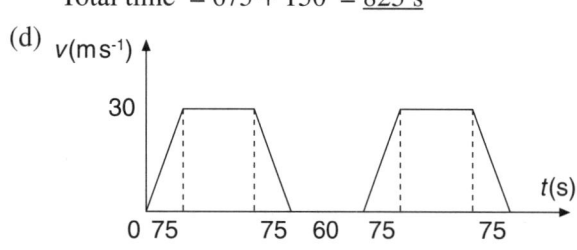

Time $= 60 + (4 \times 75) + \dfrac{(22500 - 4 \times 1125)}{30}$ **2**

$= \underline{960\,\text{s}}$ **1**

A diagram helps to clarify your thoughts.

Area of each of the four triangles is 1125.

The total distance travelled must still be 22 500 m.

6 (a)   **2**   *Take t = 0, 1, 2, 3, 4.*

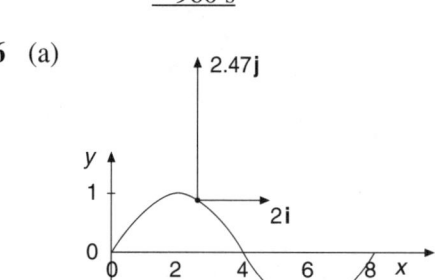

(b) $\mathbf{v} = 2\mathbf{i} + \dfrac{\pi}{2}\cos\dfrac{\pi t}{2}\,\mathbf{j}$ **1**   *Differentiating $\mathbf{r}$ w.r.t. time.*

When $t = 3$, $\underline{\mathbf{v} = 2\mathbf{i}}$ **1**

$\mathbf{a} = -\left(\dfrac{\pi}{2}\right)^2 \sin\left(\dfrac{\pi t}{2}\right)\mathbf{j}$ **1**   *Differentiating $\mathbf{v}$ w.r.t. time.*

When $t = 3$, $\underline{\mathbf{a} = 2.47\mathbf{j}}$ **1**

(See diagram)   **2**

7 Distance $= (30 \times 20) \div 2$ **1**   *Distance = area of triangle.*

$= \underline{300\,\text{m}}$ **1**

Distance travelled = Area under graph

= Area under curve + (20×20) + 300

Area under curve > $(20 \times 10) \div 2 + 700$ **1**

Distance travelled > 800 m   **1**

Area under curve > area of triangle.

Find t value of point on curve where

gradient is 2, i.e. where tangent is   **1**

parallel to line joining O to (10, 20)   **1**   *As this line has gradient 2.*

Answer	Mark	Examiner's tip

8 (a) g is constant; air resistance is negligible; ball is a particle; spin effects | 2 | For any two of these.

(b) Ball bounces when height = 0 | 1 |

i.e. when $5t - 5t^2 = 0$ | 1 |

$t = 1$ | 1 |

$\underline{x = 8.66\,\text{m}}$ | 1 |

(c) $\mathbf{v} = \begin{pmatrix} 8.66 \\ 5-10t \end{pmatrix}$ | 2 | Differentiating **r** w.r.t. time.

Speed $= \sqrt{8.66^2 + (-5)^2}$ | 1 |

$= 10\,\text{m s}^{-1}$ | 1 |

9 $64 = 2u + \frac{1}{2} \times 4a$ | 1 | Using $s = ut + \frac{1}{2}at^2$ from O to A

$250 = 5u + \frac{1}{2} \times 25a$ | 1 | and from O to B; note that we can't assume that $u = 0$.

(a) Solving to give $\underline{a = 12\,\text{m s}^{-2}}$ | 2 |

$\Rightarrow u = 20\,\text{m s}^{-1}$ | 1 |

(b) $v = 20 + 5 \times 12$ | 1 | Using $v = u + at$.

$\underline{= 80\,\text{m s}^{-1}}$ | 1 |

10 (a) $v = 3t^2 - 18t + C$ | 2 | Integrating a w.r.t. time.

$C = 15$ | 1 | Since $v = 15$ when $t = 0$.

$v = 0 \Rightarrow 3t^2 - 18t + 15 = 0$ | 1 | Divide both sides by 3 and solve.

$\underline{t = 1 \text{ or } 5\,\text{s}}$ | 2 |

(b) $s = t^3 - 9t^2 + 15t + C$ | 2 | Integrating v w.r.t. time; note $C = 0$.

When $t = 1$, $s = 7$ and $t = 5$, $s = -25$ | 2 | Using answers from (a).

Distance $= 7 - (-25) \underline{= 32\,\text{m}}$ | 2 |

11 (a) When $t = 2$, $v = 0$ | | 'Aircraft is brought to rest in 2 sec'.

$0 = u\cos 2\omega - 2k$ | 1 |

$k = \frac{1}{2}u\cos 2\omega$ | 1 |

(b) $2 = \frac{1}{2} \times 32\cos 2\omega \Rightarrow \cos 2\omega = \frac{1}{8}$ | 1 |

$\underline{\omega = 0.7227}$ | 1 |

(c) (i)

t	0.0	0.4	0.8	1.2	1.6	2.0
v	32	29.87	25.20	18.30	9.69	0.00

| 2 | Using $v = 32\cos(0.7227t) - 2t$.

(ii) Distance $=$ area under v-t graph $=$ | 1 |

$\frac{0.4}{2}\{32 + 0 + 2(29.87 + 25.20 + 18.30 + 9.69)\}$ | 2 | Using Trapezium Rule.

$\underline{= 40\,\text{m}}$ | 1 | To nearest metre.

Answer	Mark	Examiner's tip
12 (i) <u>16 cm</u>	1	When $t = 0$, $\mathbf{r}_A = 16\mathbf{j}$.
(ii) When $t = 2$, $\mathbf{r}_A = 20\mathbf{i}$, so it is on the table.	2	Need $\mathbf{j}$-component to be zero.
Thus distance is <u>20 cm.</u>	1	
(iii) $\mathbf{v}_A = 10t\,\mathbf{i} - 8t\,\mathbf{j}$	1	Differentiating $\mathbf{r}_A$ w.r.t. time.
<u>$t = 0 : 0, 0 ; t = 2 : 20, -16$</u>	2	Taking $\mathbf{i}$ and $\mathbf{j}$ components in each case.
(iv) When $t = 0$, $\mathbf{r}_A = 16\mathbf{j} = \mathbf{r}_B$	1	
When $t = 2$, $\mathbf{r}_A = 20\,\mathbf{i} = \mathbf{r}_B$	1	
(v) $\mathbf{v}_B = (30t - 15t^2)\mathbf{i} + (-48t + 48t^2 - 12t^3)\mathbf{j}$	2	Differentiating $\mathbf{r}_B$ w.r.t. time.
When $t = 0$, $\mathbf{v}_B = 0$; when $t = 2$, $\mathbf{v}_B = 0$	1	
Model B as vertical component of velocity is zero when $t = 2$ and		Whereas $\mathbf{v}_A$ has vertical component -16 when $t = 2$, so chip would strike bench hard.
'... is to be placed *gently*....'	2	

3 STATICS OF A PARTICLE

Answer	Mark	Examiner's tip
1 (a)		A clear, labelled diagram is essential as this defines your symbols.

Answer	Mark	Examiner's tip
For A,		It is important to say which body you are considering.
$N = 2g \cos\theta$	2	Resolving perpendicular to slope.
$\quad = \underline{1.6g}$	1	
(b) For B, $T = 2.2g$	1	Resolving vertically.
For A, $T = F + 2g \sin\theta$	2	Resolving parallel to slope.
$2.2g = F + 1.2g$	1	
$F = 1g$	1	
$1g = \mu\,1.6g$	1	Limiting equilibrium $\Rightarrow F = \mu N$.
$\underline{\mu = 0.625}$	1	
2 The new force must balance the 2 N force	1	
Hence a force of magnitude 2 N,	1	
in the direction opposite to the 2 N force	1	

Answer	Mark	Examiner's tip

3

$$\mathbf{W} = \begin{pmatrix} 0 \\ 0 \\ -5g \end{pmatrix}$$

1   A mass of 5 kg has weight of $5g$ N vertically downwards.

$\mathbf{T}_1 + \mathbf{T}_2 + \mathbf{T}_3 + \mathbf{W} = 0$

1   Light is in equilibrium.

$\mathbf{T}_3 = -\mathbf{T}_1 - \mathbf{T}_2 - \mathbf{W}$

1   Rearranging.

$$= \begin{pmatrix} 80 \\ 0 \\ 29 \end{pmatrix} \text{ or } \begin{pmatrix} 80 \\ 0 \\ 30 \end{pmatrix}$$

1   Use $g = 9.8$ or $10\,\text{m s}^{-2}$.

4 (i) No horizontal force to balance the horizontal component of tension

1
1

(ii)

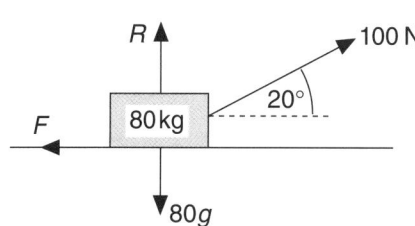

2   Clear, labelled diagram.

(iii) $F = 100 \cos 20$

1   Resolving horizontally.

$\quad\quad = 94.0\,\text{N}$

1   3 s.f. (Make sure that your calculator is in degrees mode!).

$\quad R = 80g - 100 \sin 20$

2   Resolving vertically.

$\quad\quad = 750\,\text{N}$

1   3 s.f. (use $g = 9.8\,\text{m s}^{-2}$).

(iv) $T \cos 20 = 120$

1   Resolving horizontally.

$\quad T = 128\,\text{N}$

1   3 s.f.

(v) $140 \cos 20 - 120 = 80a$

2   Newton's 2nd Law.

$\quad a = 0.144\,\text{m s}^{-2}$

1   3 s.f.

5 (i) $\arctan(\frac{1}{4}) = 14°$; $\arctan(\frac{1}{6}) = 9.46°$

1

(ii) $400\,\text{N}$

1   Since pulley is modelled as being small and light.

(iii)

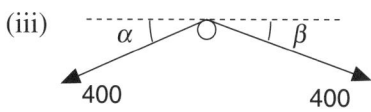

2

(iv) $400(\cos \beta - \cos \alpha) = 6.5\,\text{N}$

3   Resolving horizontally $\rightarrow$.

$400(\sin \beta + \sin \alpha) = 162.8\,\text{N}$

2   Resolving vertically $\downarrow$.

Hence tension in wire has components $6.5\,\text{N} \leftarrow$ and $162.8\,\text{N} \uparrow$

2

(v)

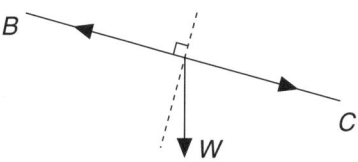

1   Resolving perpendicular to the wire,

1   weight of washing has component

1   but tension does not, hence wire cannot remain straight.

Answer	Mark	Examiner's tip

6

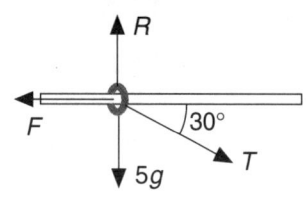

$F = \frac{1}{2}R$ | **1** | Since equilibrium is limiting, $F = \mu R$.

$F = T\cos 30°$ | **1** | |

$R = 5g + T\cos 60°$ | **2** | Resolving vertically $g = 9.8\,\mathrm{m\,s}^{-2}$.

$T\cos 30° = \frac{1}{2}(5g + T\cos 60°)$ | **1** | Eliminating R.

$\underline{T = 40\,\mathrm{N}}$ | **1** | To 2 s.f.

7

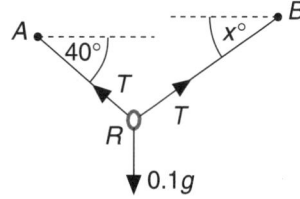

| | **1** | Note that as the ring is smooth the tensions in both sections of the string are equal. |

$T\cos 40° = T\cos x° \Rightarrow x = 40$ | **1** | Resolving horizontally.

$2T\sin 40° = 0.1g$ | **1** | Resolving vertically. $g = 9.8\,\mathrm{m\,s}^{-2}$

$\underline{T = 0.76\,\mathrm{N}}$ | **1** | 2 s.f.

8

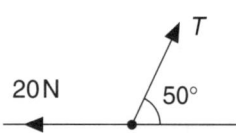

| | | A clear, labelled diagram is essential. |

$T\cos 50° - 20 = 0$ | **2** | Since the suitcase is moving with constant speed, it is in equilibrium.

$\underline{T = 31\,\mathrm{N}}$ | **1** | 2 s.f. for this exam board as stated in rubric on front of paper.

9

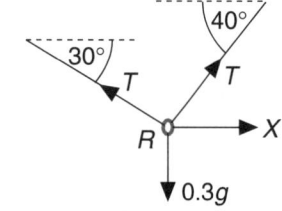

| | | Don't forget to include the weight of the ring on your diagram. Note also that as the ring is smooth the tension in both parts of the string is the same. |

$T\sin 30° + T\sin 40° - 0.3g = 0$ | **2** | Resolving vertically.

$T(\sin 30° + \sin 40°) = 0.3 \times 9.81$ | | |

$\underline{T = 2.6\,\mathrm{N}}$ | **1** | To 2 s.f.

$X + T\cos 40° - T\cos 30° = 0$ | **2** | Resolving horizontally.

$X = T(\cos 30° - \cos 40°)$ | | In subsequent calculations always use unrounded values.

$\underline{X = 0.26\,\mathrm{N}}$ | **1** | To 2 s.f.

4 DYNAMICS OF A PARTICLE

Answer	Mark	Examiner's tip

1 (a) Resistance = driving force

As she is running at constant speed.

= 120 ÷ 4 — 1 — Using Power = $Fv \Rightarrow F = P \div v$.

= 30 N — 1

(b)

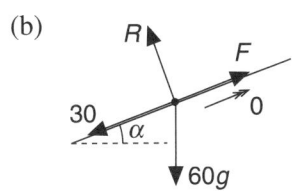

A simple diagram helps to clarify the situation and reminds you to include the weight of the woman in your calculations.

$F - 30 - 60g \times \frac{1}{15} = 0$ — 2 — Resolving up the slope.

$F = 69.2$ N

Power $= 69.2 \times 3$ — 1 — Again using Power = Fv.

$= 207.6$ W — 1

(c) When $v = 4$, $R = 30$

From part (a).

If R is proportional to v^2, then, when $v = 3$,

$R = (\frac{3}{4})^2 \times 30$ — 1

$= 16.875$ N — 1

(d) New driving force $= 16.875 + 60g \times \frac{1}{15}$ — 1

New power $= (16.875 + 60g \times \frac{1}{15}) \times 3$ — 1

$= 168.2$ W — 1

2

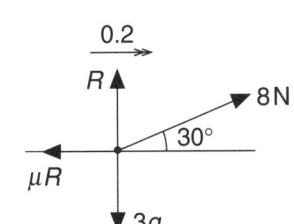

A clear diagram, showing all the forces *and* the acceleration, is again essential. Note that as the particle is moving friction will be limiting and this is best included on the diagram. A common error, in situations like this, is to assume that the normal contact force is equal to the weight of the particle.

$R + 8 \sin 30 - 3g = 0$ — 2 — Resolving vertically – note that we must resolve the acceleration as

$R = 3 \times 9.81 - 4 = 25$ N (2 s.f.) — 1

$8 \cos 30 - \mu R = 3 \times 0.2$ — 2 — Resolving horizontally.

$\mu = 0.25$ (2 s.f.) — 1 — Use the exact not the rounded value for R.

3 (a) (i)   (ii) — 2

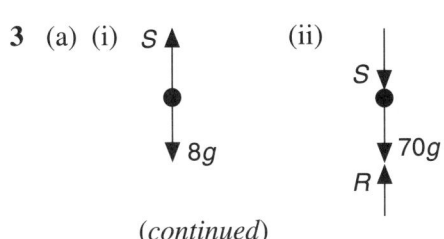

(*continued*)

Answer	Mark	Examiner's tip

(b) For the suitcase, $S - 8g = 8 \times 2$ **1** Resolving in the direction of the
$\underline{S = 96 \text{ N}}$ **1** acceleration. ($g = 10 \,\text{m s}^{-2}$)

For the man, $R - S - 70g = 70 \times 2$ **1** Resolving in the direction of the
$\underline{R = 936 \text{ N}}$ **1** acceleration.

4

A simple diagram is all that is required.

250 ← [12 000] → 400 (with arrow a above)

(i) $400 - 250 = 12\,000a$ **3** Resolving in direction of
$\underline{a = 0.0125 \,\text{m s}^{-2}}$ **1** acceleration.

(ii) $v^2 = 2 \times .0125 \times 10$ **1** Since a is constant, we can use
$v = 0.5 \,\text{m s}^{-1}$ $v^2 = u^2 + 2as$, with $u = 0$.
$0.5 = 0.0125t$ **1** Using $v = u + at$.
$\underline{t = 40 \text{ s}}$ **1**

(iii) (A)

It is best to draw a fresh diagram; remember to include the acceleration.

150 ← [12 000] ← 250 (with arrow a above)

$-250 - 150 = 12\,000a$ **1** Resolving in direction of
acceleration.

$a = -\frac{1}{30} \,\text{m s}^{-2}$ **1** Note that a is negative because of the way in which it was marked on the diagram.

$0 = 0.5 + (-\frac{1}{30})t \Rightarrow t = 15 \text{ s}$ **1** $v = u + at$ with $v = 0$.

(B) Distance travelled whilst stopping
$= \frac{0.5 + 0}{2} \times 15 = 3.75 \text{ m}$ **1** Using $s = \frac{(u+v)t}{2}$.

Total distance $= 3.75 + 10 = \underline{13.75 \text{ m}}$ **1**

(C) Total time $= 40 + 15 = \underline{55 \text{ s}}$ **1**

5 (i) Resistance $=$ driving force

Steady speed $\Rightarrow$ no acceleration.

$= 40\,000 \div 25$ **1** $P = Fv \Rightarrow F = P \div v$
$= \underline{1600 \text{ N}}$ **1**

(ii)

It is worth drawing a diagram here.

$F - 1600 - 800g \sin\theta = 0$ **2** Resolving up the slope.

$F = 2384 \text{ N}$ **1** $g = 9.8 \,\text{m s}^{-2}$ and $\sin\theta = \frac{1}{10}$.

New $P = 2384 \times 25 = 59\,600 \text{ W}$ **1** $P = Fv$.

Extra power $= 59\,600 - 40\,000 = \underline{19\,600 \text{ W}}$ **1**

Answer	Mark	Examiner's tip

(iii) Work done against resistance = $900x$ — 1 — Where x is distance travelled.

K.E. loss = $\frac{1}{2} \times 800 \times 25^2$ — 1

P.E. gain = $800g\, x \sin\theta$ — 2 — Vertical gain in height is $x \sin\theta$.

Work done = K.E. loss − P.E. gain

$900x = \frac{1}{2} \times 800 \times 25^2 - 800g\, x \sin\theta$ — 2

$\underline{x = 148.5\ \text{m}}$ — 1 — Note that this problem could also be solved by resolving up the slope to find the deceleration then using $v^2 = u^2 + 2as$ with $v = 0$.

6 (a) (i) $\quad I = mv - mu$ — It is important to choose a positive direction since I, v and u are all vectors.

$= 0.15\,(40 - (-25))$ — 3

$\underline{= 9.75\ \text{N s}}$ — 1

(ii) $\quad 9.75 = F \times 0.01$ — 2 — Using Impulse = force x time.

$\underline{F = 975\ \text{N}}$ — 1

(b) $1.5 \times 1 + 1.2 \times 0 = 1.5v + 1.2 \times 0.8$ — 4 — Momentum is conserved.

$\underline{v = 0.36\ \text{m s}^{-1}}$ — 1

7 (a) Angle of slope — 1 — Note that area of surface in contact is not a factor.

Magnitude of normal reaction — 1

Nature of surfaces in contact — 1

(b) (i) $\quad mg \sin 30° = ma$ — 1 — Resolving down the slope.

$\underline{a = 5\ \text{m s}^{-2}}$ — 1 — $g = 10\ \text{m s}^{-2}$ in this question.

(ii) $\quad 20 = \frac{1}{2} \times 5t^2$ — 1 — As a is constant we can use $s = ut + \frac{1}{2}at^2$.

$t = \sqrt{8} = \underline{2.83\ \text{s}}$ — 1

(c) $20 = 2 \times 4 + \frac{1}{2}a \times 4^2$ — 1 — Using $s = ut + \frac{1}{2}at^2$ again.

$\underline{a = 1.5\ \text{m s}^{-2}}$ — 1

(d) — A diagram is essential here – note that friction is limiting as there is motion so $F = \mu R$. This should be included on the diagram.

$R = mg \cos 30°$ — 1 — Resolving perpendicular to slope.

$mg \sin 30° - \mu R = 1.5m$ — 2 — Resolving down slope with $a = 1.5$ from part (c).

$5m - \mu \times 10m \cos 30° = 1.5m$ — Substituting for R and solving to find μ.

$\underline{\mu = 0.404}$ — 1

(e) $\underline{\text{Air resistance}}$ — 1

8 $2.1V = 0.1 \times 420$ — 2 — Momentum is conserved.

$\underline{V = 20\ \text{m s}^{-1}}$ — 1

Answer	Mark	Examiner's tip
9 $15\,000 = F \times \dfrac{120\,000}{3600} = F \times \dfrac{100}{3}$	2	Power = force × speed where speed must be in m s^{-1}.
		Max speed $\Rightarrow$ acceleration = 0
$F = 450\,\text{N}$		
$R = 450\,\text{N}$	1	$\Rightarrow$ resistance = driving force.
New $F = 15\,000 \div (\frac{50}{3}) = 900\,\text{N}$	2	F has doubled since v has halved.
$900 - 450 = 800a$	1	Resolving in direction of acceleration.
$\underline{a = 0.56\,\text{m s}^{-2}}$	1	To 2 s.f.

10 (a)

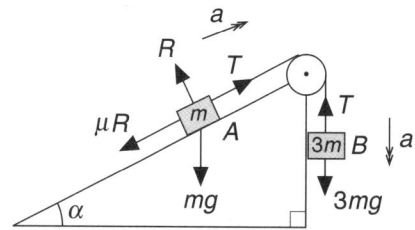

	Mark	Examiner's tip
	1	A clear diagram showing forces and accelerations is essential. As system is in motion, $F = \mu R$.
For A, $\qquad R = mg\cos\alpha = 0.8mg$	2	Resolving perpendicular to slope.
$T - \mu R - mg\sin\alpha = ma$	3	Resolving up slope, $\mu = \frac{3}{4}$,
i.e. $\ T - \frac{3}{4} \times 0.8mg - 0.6mg = ma$		$R = 0.8mg$ and $\sin\alpha = 0.6$.
i.e. $\qquad\qquad T - 1.2mg = ma$		Always simplify your equations.
For B, $\quad 3mg - T = 3ma$	2	Resolving vertically downwards.
$3mg - 1.2mg = 4ma$	1	Adding the two equations to eliminate T.
$\underline{a = 0.45g\,\text{m s}^{-2}}$	1	
(b) $v^2 = 2 \times 0.45g \times 1 = 0.9g$	2	v is speed of A just as rope goes slack when B hits the ground. Note that we need v^2 in equation below so there is no need to square root.
For A, after rope goes slack,	2	This is same equation as above with $T = 0$ since rope is now slack.
$-1.2mg = ma_1$		
$a_1 = -1.2g$	1	a_1 is negative as A decelerates as it continues to slide up the plane.
$0^2 = 0.9g + 2(-1.2g)s$	1	Using $v^2 = u^2 + 2as$.
$s = 0.375\,\text{m}$	1	Note that g cancels.
$\underline{\text{Total distance} = 1 + 0.375 = 1.375\,\text{m}}$	1	Adding in the initial 1 m moved before B hits the ground.
11 (a) $\quad 900 \times 7 = (900 + 500)\,v$	2	Momentum before = Momentum after.
$\underline{v = 4.5\,\text{m s}^{-1}}$	1	
(b) For B, $\ I = 500 \times 4.5$	1	To find the impulse we need to consider just one of the trucks – it is easier to consider B as it is initially at rest.
$\underline{I = 2250\,\text{N s}}$	1	

Answer	Mark	Examiner's tip

(c) K.E. Loss = Total K.E. Before − Total K.E. After

$= (\frac{1}{2} \times 900 \times 7^2) - (\frac{1}{2} \times 1400 \times 4.5^2)$ **3**

$\underline{= 7875 \text{ J}}$ **1**

Here we must consider both trucks.

(d) Work done against buffers = K.E. Loss

$95\,000 \times d = \frac{1}{2} \times 1400 \times 4.5^2$ **3**

$\underline{d = 0.15 \text{ m}}$ **1**

12 Distance moved = $2 \times 10 = 20$ m

Work done = $25 \cos 30° \times 20$ **2**

$\underline{430 \text{ J to 2 s.f.}}$ **1**

Note that we need the horizontal component only of the force as the vertical component does no work.

13 Work done against friction = K.E. Loss

$\mu \times 1g \times 2 = \frac{1}{2} \times 1 \times (10^2 - 8^2)$ **3**

$\underline{\mu = 0.92 \text{ to 2 s.f.}}$ **1**

$6 = \frac{1}{2} \times 9.81 \times t^2$ **1**

$t = 1.106$ s **1**

$\underline{s = 8 \times 1.106 = 8.8 \text{ m to 2 s.f.}}$ **1**

Boat will be moving at 45° to horizontal when horizontal and vertical components of velocity are equal, i.e. both are $8\,\text{ms}^{-1}$. **1**

$8^2 = 0^2 + 2 \times 9.81 \times s$ **2**

$s = 3.262$ m **1**

$\underline{\text{Height above water} = 6 - 3.262 = 2.7\,\text{m}}$ **1**

Note that $R = 1g$.

$s = ut + \frac{1}{2}at^2$ vertically with $u = 0$.

$s = ut$ horizontally as $a = 0$.

Since horizontal component is 8 ms^{-1} throughout the motion.

$v^2 = u^2 + 2as$ vertically.

To 2 s.f.

5 FURTHER MOMENTUM

Answer	Mark	Examiner's tip

1

masses $\quad$ m $\quad$ m

speeds: before $\quad \longrightarrow 2 \quad \longrightarrow 1$

after $\quad \longrightarrow v_1 \quad \longrightarrow v_2$

Draw a clearly labelled diagram showing masses and speeds before and after impact. Ensure that the $2\,\text{ms}^{-1}$ and $1\,\text{ms}^{-1}$ are the correct way round so that collision occurs.

Conservation of linear momentum: **1**

$2m + 1m = mv_1 + mv_2$

$\therefore \quad 3 = v_1 + v_2$

(*continued*)

Use and simplify the C.L.M. equation.

Answer	Mark	Examiner's tip

Restitution Equation (Newton's Experimental Law):

$0.8(2 - 1) = v_2 - v_1$ — **1** — Use and simplify N.E.L.

$\therefore \quad 0.8 = v_2 - v_1$

$\therefore \quad 2v_1 = 2.2 \Rightarrow v_1 = 1.1, v_2 = 1.9$ — **1, 1** — Solve the simultaneous equations and check that the answer makes sense. i.e: $v_2 > v_1$.

2 (a) Consider motion under gravity — Use constant acceleration equations.

$u = 0, s = 0.6, a = 9.8, v = ?$ — Write down known values.

Use $v^2 = u^2 + 2as$ — Select this equation.

$v^2 = 2 \times 9.8 \times 0.6$ — N.B. $g = 9.8\,\mathrm{m\,s^{-2}}$ for this examination.

$v = 3.43 = \sqrt{1.2g}$ — **1**

(b) Consider motion after impact — Note, as motion is upward the acceleration opposes the motion and is negative.

$v = 0, s = 0.15, a = -9.8, u = ?$

$0 = u^2 - 2 \times 9.8 \times 0.15$

$u = 1.71 = \sqrt{0.3g}$ — **1** — Similar working to (a).

Now use $u = ev$ — **1** — Use the restitution equation, noting that v is speed before impact and u is speed after impact.

$\therefore \quad e = \dfrac{1}{2}$ — **1**

3 (i)

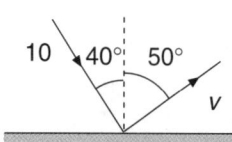

Define v as the speed after impact by a diagram or statement.

Conservation of momentum along the table:

$m \times 10 \sin 40 = m \times v \sin 50$ — **1** — Resolve velocity in each case.

$v = \dfrac{10 \sin 40}{\sin 50} = 8.39$ — **2** — Solve to find v.

(ii) Use impulse = change in momentum — The momentum before and after the impact are in different directions, $\therefore$ one is negative. The mass must be converted to kg to give units in Ns.

$|\text{Impulse}| = 2.5 \times 10^{-3}(v \cos 50 + 10 \cos 40)$ — **2**

$= 2.5 \times 10^{-3} \times 13.05$

$= 0.033\,\mathrm{Ns}$ — **2**

4 (a)

Draw a diagram showing masses and speeds.

Note that the velocity of Q is negative prior to impact.

Answer	Mark	Examiner's tip

Conservation of Linear Momentum: (C.L.M.)

$$3mu + (-2mu) = 3mv + mw$$
$$u = 3v + w \qquad \textbf{2}$$

Newton's Experimental Law: (N.E.L.)
$$e(u - (-2u)) = -v + w$$
$$\therefore \quad 3ue = -v + w \qquad \textbf{2}$$

Solving $\quad v = \dfrac{u}{4}(1 - 3e) \qquad$ **2**

(b) $v < 0 \Rightarrow 1 - 3e < 0, \therefore e > \dfrac{1}{3}$ **2** If the direction of P's motion has changed then $v < 0$ to be consistent with the diagram.

(c) $w = \dfrac{u}{4}(1 + 9e)$ **2** Solve equations for C.L.M. and N.E.L.

(d) $w_{max} = \dfrac{u}{4}(1 + 9) = \dfrac{5u}{2}$ **2** As $0 \le e \le 1$ the maximum value for $e = 1$.

(e) $e = \dfrac{5}{9}, \therefore w = \dfrac{3u}{2}, v = \dfrac{-u}{6}$ **3** Substitute the given value for e to find v and w.

After hitting the wall Q has velocity $\dfrac{-3ue'}{2}$ **1** Restitution equation for impact at the wall.

$\therefore \dfrac{3ue'}{2} > \dfrac{u}{6}$ for collision $\therefore e' > \dfrac{1}{9}$ **1, 1** $|v_Q| > |v_P|$ if Q is to catch P.

5 (i) Conservation of Linear Momentum:
$$Mu + \dfrac{1}{4}M0 = (M + \dfrac{1}{4}M)v \qquad \textbf{1}$$ Let v be the speed after impact.

$$\therefore \qquad \mathbf{v} = \dfrac{4}{5}u\mathbf{i}\ \text{m s}^{-1} \qquad \textbf{1}$$ As velocity is required, the answer is a vector.

(ii) $Mu = Mv' + \dfrac{1}{4}Mu$ **1, 1** Again total momentum is conserved, so equals initial momentum.

$$\mathbf{v'} = \dfrac{3}{4}u\mathbf{i} \qquad \textbf{1}$$

(iii)

$Mu + \dfrac{3}{4}Mu = Mv_B + Mv_A \therefore \dfrac{7u}{4} = v_B + v_A$ **1, 1** Conservation of Linear Momentum. The equations should be simplified before solving.

$\dfrac{1}{2}\left(u - \dfrac{3}{4}u\right) = v_A - v_B \therefore \dfrac{u}{8} = v_A - v_B$ **1, 1** Restitution Equation.

$\mathbf{v_A} = \dfrac{15}{16}u\mathbf{i}, \mathbf{v_B} = \dfrac{13}{16}u\mathbf{i}$ **1** Solving to obtain velocities.

(*continued*)

Answer	Mark	Examiner's tip

$$\text{Impulse} = M\left(\frac{13}{16}u - u\right)\mathbf{i}$$

1 — Impulse is a vector and the direction must be stated or implied.

$$= \frac{-3Mu}{16}\mathbf{i}\,\text{Ns}$$

1

(iv) Slows as insect starts.

1 — For either of these comments the first mark is earned.

Returns to $\frac{4}{5}u$ as insect stops.

Stays constant while insect walks.

1 — This comment earns the second mark.

6 (i)

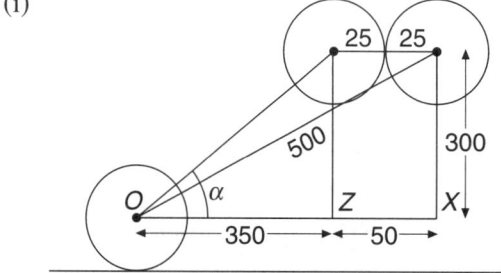

1 — Show line of centres parallel to the cushion.

(ii) $\quad OX = 400\ (3, 4, 5\ \Delta)$

1 — Use geometry and trigonometry to find the angle required.

$\therefore\ OZ = 350\ (400 - 2 \times 25)$

1

$\therefore \tan\alpha = \dfrac{300}{350} \Rightarrow \alpha = 40.6°$

1

(iii)

— Consider the component velocities along the line of centres and perpendicular to the line of centres.

1 — Use C.L.M. perpendicular to the line of centres for each ball.

Conservation of Linear Momentum:

$mW\cos 40.6 = mX + mY$

1 — Use C.L.M. along line of centres at impact.

Newton's Experimental Law:

$0.9W\cos 40.6 = Y - X$

1 — Use Restitution Equation.

$2Y = 200 = W(\cos 40.6 + 0.9\cos 40.6)$

1

$\underline{W = 138.6\ \text{mm s}^{-1}}$

1 — Solve simultaneous equations to obtain W.

6 LINEAR MOTION UNDER A VARIABLE FORCE

Answer	Mark	Examiner's tip
1 $v\dfrac{dv}{dx} = 10 - 0.5v^2$	2	Here we must use $v\dfrac{dv}{dx}$ rather than
$\dfrac{v\,dv}{10 - 0.5v^2} = dx$	2	$\dfrac{dv}{dt}$ for a since we need to find x.
$\displaystyle\int_0^2 \dfrac{v\,dv}{10 - 0.5v^2} = \int_0^x dx$	3	Or you could use indefinite integrals with an arbitrary constant rather than limits.
$\left[-\ln(10 - 0.5v^2)\right]_0^2 = \left[x\right]_0^x$	2	
$\underline{x = \ln\left(\tfrac{5}{4}\right) = 0.223\ \text{m}}$	1	
2 (a) (i) <u>Zero</u>	1	
(ii) <u>Non-zero</u>	1	
<u>At right angles to tangent to bend</u>	1	
(b) At top speed $a = 0 \Rightarrow F = R$	1	Where F is driving force.
Let $R = kv$		By assumption.
$600 = F \times 10 \Rightarrow F = 60\ \text{N} \Rightarrow R = 60\ \text{N}$	1	Using $P = Fv$.
so $60 = k \times 10 \Rightarrow k = 6$		
<u>Hence $R = 6v$ is model.</u>	1	
(c) If F is constant then $F = 60$ from above	1	
$60 - 6v = 90\dfrac{dv}{dt}$	1	Resolving horizontally.
$\dfrac{dv}{dt} = \dfrac{10 - v}{15}$	1	
(d) $\displaystyle\int \dfrac{15}{10 - v}\,dv = \int dt$	1	Separating the variables.
$-15\ln(10 - v) = t + C$	1	
When $t = 0,\ v = 0 \Rightarrow C = -15\ln 10$	1	
$-t = 15\ln(10 - v) - 15\ln 10$		
$-\dfrac{t}{15} = \ln\left(\dfrac{10 - v}{10}\right)$	1	Combine the log terms before raising both sides to power e.
$10e^{-t/15} = 10 - v$		
$v = 10\left(1 - e^{-t/15}\right)$	1	
(e) Air resistance is not proportional to v.	1	Normally proportional to v^3.
3 $-\dfrac{km}{x^2} = mv\dfrac{dx}{dx}$	3	Any derivative form of acceleration has its positive direction in the direction of x increasing – hence the negative sign in this equation.
(continued)		

Answer	Mark	Examiner's tip
$\int -\left(\dfrac{k}{x^2}\right) dx = \int v\,dv$	1	Separating the variables.
$\dfrac{k}{x} = \dfrac{1}{2}v^2 + C$	2	Integrating both sides.
When $x = a$, $v = u \Rightarrow C = \dfrac{k}{a} - \dfrac{1}{2}u^2$	1	
$v = \sqrt{u^2 + 2k\left(\dfrac{1}{x} - \dfrac{1}{a}\right)}$	1	
If $u^2 - 2\dfrac{k}{a} > 0$, then $v > 0$ for all x as $k > 0$ and $x > 0$	2	Hence rocket is always moving away from the Earth.

4 (a) (i) $-F = m\dfrac{dv}{dt}$, $-kmg = m\dfrac{dv}{dt}$ **1**   Using Newton's Second Law.

$\dfrac{dv}{dt} = -kg$ **1**

(ii) $v = -kgt + C$ **1**   Integrating w.r.t. time.

$t = 0$, $v = u \Rightarrow C = u$ **1**   Using initial conditions.

Hence $\underline{v = u - kgt}$

(b) (i) $-F - D = m\dfrac{dv}{dt}$,

$-kmg - bmv = m\dfrac{dv}{dt}$ **1**   Using Newton's Second Law.

$\dfrac{dv}{dt} = -bv - kg$ **1**

(ii) $-\int \dfrac{dv}{bv + kg} = \int dt$ **1**   It is easier to take out the negative sign.

$-\dfrac{1}{b}\ln(bv + kg) = t + B$ **1**   Integrating both sides.

$t = 0$, $v = u \Rightarrow B = -\dfrac{1}{b}\ln(bu + kg)$ **1**   Using initial conditions.

$t = \dfrac{1}{b}[\ln(bu + kg) - \ln(bv + kg)]$

$t = \dfrac{1}{b}\ln\left[\dfrac{bu + kg}{bv + kg}\right]$ **1**   Collecting log terms.

(c) $0 = 30 - 0.7 \times 10 \times t_1 \Rightarrow t_1 = 4.29$ s   **1**   Using result from (a)(ii).

$t_2 = \dfrac{1}{0.2}\ln\dfrac{(0.2 \times 30 + 0.7 \times 10)}{(0 + 0.7 \times 10)}$ Using result from (b).

$t_2 = 3.10$ s   **1**

$\underline{t_1 - t_2 = 1.19 \text{ s}}$ **1**

7 MOMENTS, CENTRES OF MASS AND EQUILIBRIUM

Answer	Mark	Examiner's tip

1

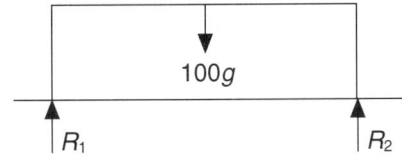

Resolve vertically: $R_1 = R_2 = 50g = 490 \text{ N}$ **1, 1**

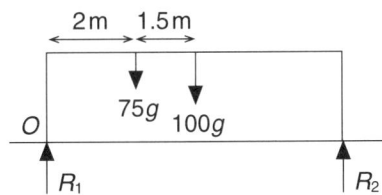

$\circlearrowleft O$: $75g \times 2 + 100g \times 3.5 = R_2 \times 7$ **1**

$\qquad\qquad 500g = 7R_2$

$\qquad\qquad \underline{R_2 \quad = 700 \text{ N}}$ **1**

Resolve vertically: $R_1 + R_2 = 175g$ **1**

$\qquad\qquad \therefore R_1 = 1015 \text{ N}$ **1**

Draw a diagram showing the forces acting. Symmetry is used to answer this question. $g = 9.8 \, \text{m s}^{-2}$

Draw a new diagram.

Take moments about the foot of one of the posts to eliminate one of the unknown forces.

Could check answers by taking moments about the foot of the other post.

2

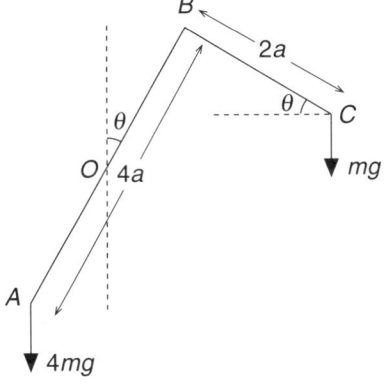

$\circlearrowleft O$:   **1**

$4mg \times 2a \sin\theta = mg \times (2a\cos\theta + 2a\sin\theta)$ **1, 1**

$\therefore \quad 8\sin\theta = 2\cos\theta + 2\sin\theta$

$\therefore \quad 6\dfrac{\sin\theta}{\cos\theta} = 2\dfrac{\cos\theta}{\cos\theta}$ **1**

$\therefore \quad \tan\theta = \dfrac{1}{3} \Rightarrow \underline{\theta = 18.4^\circ}$ **1**

AB and BC are light rods so assume their mass is zero.

The forces at B are equal and opposite, so their total moments will be zero.

To eliminate the unknown forces at O take moments about O. Use right angle triangles to find distances.

Collect terms and divide by $\cos\theta$ to obtain $\tan\theta$.

Answer	Mark	Examiner's tip

3 (a) <u>The weight of the rod can be ignored.</u>

(b)

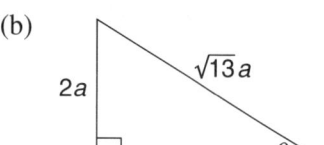

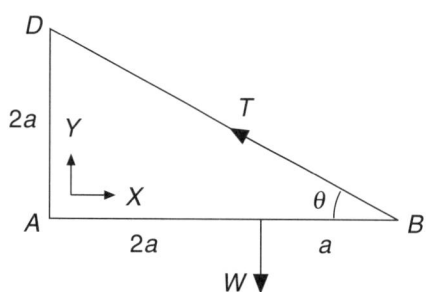

$$\circlearrowleft A: \quad W \times 2a = T\sin\theta \times 3a$$

$$T = \frac{2W}{3\sin\theta} = \frac{W\sqrt{13}}{3}$$

(c) $(\rightarrow)\ X = T\cos\theta = \frac{W\sqrt{13}}{3} \times \frac{3}{\sqrt{13}} = \underline{W}$

$(\uparrow)\ Y + T\sin\theta = W$

$$Y = W - \frac{W\sqrt{13}}{3} \times \frac{2}{\sqrt{13}} = \underline{\frac{W}{3}}$$

4 (i)

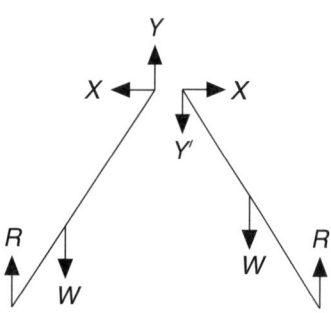

By symmetry $Y' = -Y$

Since internal forces $Y = Y'$

$\therefore Y = Y' = 0.$

(ii)

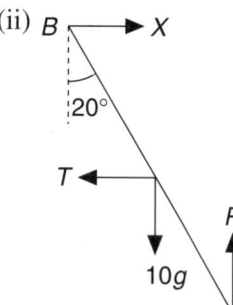

Weight and reaction forces shown

Tension and force at hinge shown

$(\rightarrow)\ X = T$

Marks column (right):

3 (a) — **1** — 'Light' rod means mass is zero.

Draw triangle and deduce from Pythagoras' theorem that the length $BD = \sqrt{13}a$.

$$\therefore \sin\theta = \frac{2}{\sqrt{13}}, \cos\theta = \frac{3}{\sqrt{13}}$$

moments — **1** — Take moments about A to eliminate X and Y.

T — **1, 1** — Use *exact values* and not approximations from your calculator to obtain the printed answer.

(c) — **1, 1** — Resolve horizontally and vertically to obtain equations in X and Y and solve to find the required forces

$Y+T\sin\theta$ — **1**

Y — **1**

By symmetry — **1** — Two marks requires both equations and a statement to complete the argument.

$\therefore Y=Y'=0$ — **1**

Weight and reaction — **1** — Clearly labelled diagrams needed, showing these forces.

Tension and force — **1**

$(\rightarrow) X = T$ — **1** — Resolve horizontally.

Answer	Mark	Examiner's tip
(iii) $\circlearrowright C$: $2X\cos 20 - T\cos 20 - 98\sin 20 = 0$	1, 1	Clear moment equation about C, or about midpoint. Use $g = 9.8\,\mathrm{m\,s^{-2}}$
$T = 98\tan 20 \approx 35.7\,\mathrm{N}$	1	Eliminate X to obtain T.

(iv)

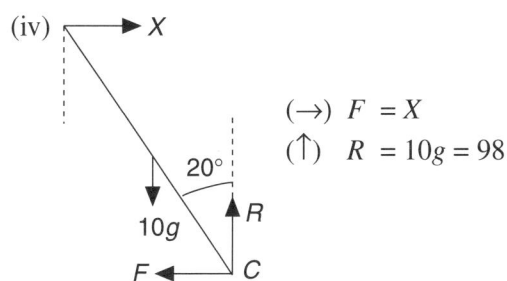

$(\rightarrow)$ $F = X$ **1**

$(\uparrow)$ $R = 10g = 98$ **1**   Three equations are needed this time.

Resolve horizontally.
Resolve vertically.

$\circlearrowright C$: $98\sin 20 = 2X\cos 20$	1	Correct moment equation.
$X = 49\tan 20 \approx 17.8$	1	Obtain X or F (these are equal).
Use $F \le \mu R$	1	Use of inequality.
$\therefore \mu \ge \dfrac{17.8}{98} \approx 0.18$	1	Obtain answer.

5 (a) <u>Light rods and smooth pin joints.</u>	1, 1	Some variation is allowed here, e.g. inextensible rods.
(b) $\circlearrowright D$: $100\,a = 2aQ \Rightarrow Q = 50\,\mathrm{N}$	2	Take moments about D to eliminate the forces at D. $2a$ is the length of the diagonal.

(c)

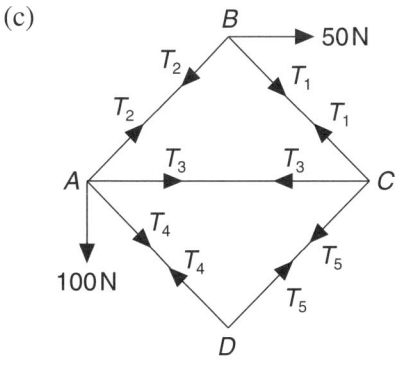

Start by assuming that all the rods are under tension and show the forces on your diagram.

At B: $(\uparrow)$ $T_1\cos 45 = -T_2\cos 45 \Rightarrow T_1 = -T_2$		Resolve vertically and horizontally at points with the least number of unknown forces.
$(\rightarrow)$ $50 + T_1\cos 45 = T_2\cos 45$	1	
$\therefore \underline{T_1 = -25\sqrt{2}\,\mathrm{N}, \ T_2 = +25\sqrt{2}\,\mathrm{N}}$	1, 1	
At A: $(\uparrow)$ $T_2\cos 45 = T_4\cos 45 + 100$		
$\therefore \underline{T_4 = -75\sqrt{2}\,\mathrm{N}}$	1	Note that T_1, T_4 and T_5 are *negative* so the rods are not experiencing tension, they are in *compression*.
$(\rightarrow)$ $T_3 + T_2\cos 45 + T_4\cos 45 = 0$		
$\therefore \underline{T_3 = -25 + 75 = 50\,\mathrm{N}}$	1	
At C: $(\uparrow)$ $T_1\cos 45 = T_5\cos 45$		
$\therefore \underline{T_5 = -25\sqrt{2}\,\mathrm{N}}$	1	
(d) <u>*AB* and *AC*.</u>	1, 1	These rods are under tension so can be replaced by ropes.

Answer	Mark	Examiner's tip

6 (a) <u>Symmetry.</u>

    (b) Area of component $= 25\pi - 4\pi$

        $\circlearrowleft A$: $21\pi\,\bar{x} = 4\pi \times 2$

                   $\bar{x} = \dfrac{8}{21}$

    (c) $kM \times 5 = M \times \dfrac{8}{21}$

      $\therefore$ $\underline{k = \dfrac{8}{105}\ \text{or}\ 0.076}$

Marks: **1**, **2**, **3**, **1**, **3**

Examiner's tips:
Centre of mass lies on a line of symmetry as object is uniform.

Equation must be dimensionally correct.

Answer printed so must follow correct working.

Again moments are taken about *A*.

7

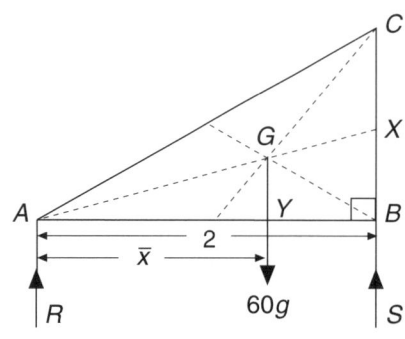

    (i) $\bar{x} = \dfrac{2}{3} \times 2m = \dfrac{4}{3}\,m$ **1**

Use similar triangles *AGY* and *AXB* and median property:

$$AG = \dfrac{2}{3}AX$$

    (ii) $\circlearrowright A$: $60g \times \dfrac{4}{3} = S \times 2$

Take moments about *A* to eliminate force *R*.

              $S = 40g$ **2**

      $(\uparrow)$ $R + S = 60g$

             $\underline{R = 20g}$ **2**

As the body is in equilibrium, total vertical component of forces = 0.

8 (a)

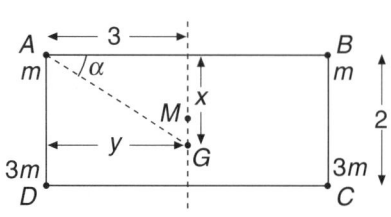

    $\underline{y = 3\,\text{cm}}$ **1**

The diagram in the question was not to scale and may confuse you. Draw your own diagram marking *G* the centre of mass. Turn the rectangle for your own convenience.
The centre of mass lies on a line of symmetry.

Answer	Mark	Examiner's tip

(b) Use $\sin \alpha = \dfrac{5}{13}$ to obtain

  $\tan \alpha = \dfrac{5}{12}$ **2**

You may use a 5, 12, 13 triangle or your calculator, but if you do use your calculator do *not* approximate the answer.

Use $\tan \alpha = \dfrac{x}{3}$ to find $\underline{x = 1.25\,\text{cm}}$ **2**

The method uses the principle that the centre of mass lies on the vertical through the point of suspension so that $\alpha = \angle BAG$.

(c) Use $\sum m_i \bar{x} = \sum m_i \bar{x}_i$

  $(8m + M) \times 1.25$

  $= 3m \times 2 + 3m \times 2 + M \times 1$ **5**

  $\therefore\ 10m + 1.25M = 12m + M$

  $\therefore\qquad 0.25M = 2m$

  $\qquad m = \dfrac{1}{8}M$ **3**

Moment of total mass = sum of moments.

Distances are measured from an axis along AB.

Solve making m the subject of the formula.

9 (i)

Shape	Weight (vol)	Distance of C of G from P	Moment about P
Cylinder	$2\pi R^3$	R	$2\pi R^4$
Cone	$\dfrac{2}{3}\pi R^3$	$2R + \dfrac{1}{4}(2R)$	$\dfrac{5}{3}\pi R^4$
Body	$\dfrac{8}{3}\pi R^3$	$\bar{x}$	$\dfrac{11}{3}\pi R^4$

A table clarifies the working. As weight is proportional to volume it is sufficient to use volumes in the moments equation.

5

  $\dfrac{8}{3}\pi R^3\, \bar{x} = \dfrac{11}{3}\pi R^4$ **3**

Moments equation is solved to find $\bar{x}$.

  $\therefore\qquad \bar{x} = \dfrac{11}{8}R$ **1**

(ii)   **2**

The diagram should show G vertically above A when the body is about to topple.

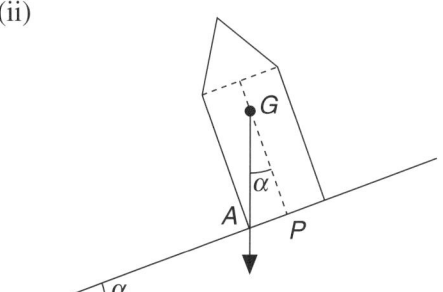

  $GP = \dfrac{11}{8}R$; $AP = R$.   **2**

  $\tan \alpha = \dfrac{R}{\frac{11}{8}R} = \dfrac{8}{11}$

  $\therefore\ \alpha = \tan^{-1}\left(\dfrac{8}{11}\right) = \underline{36.03°}$ **2**

Answer	Mark	Examiner's tip
10 (a) $$V = \pi \int_{1}^{2} y^2 \, dx$$	1	This part is pure mathematics and involves applying a standard formula.
$$= \pi \int_{1}^{2} \frac{1}{x^2} \, dx = -\pi \left[\frac{1}{x} \right]_{1}^{2}$$	1, 1	
$$\underline{V = \frac{\pi}{2}}$$	1	
(b) $$\frac{\pi}{2} \bar{x} = \pi \int_{1}^{2} xy^2 \, dx$$	1, 1	This involves using the moments equation and integration.
$$\bar{x} = 2 \int_{1}^{2} \frac{1}{x} \, dx = 2[\ln x]_{1}^{2}$$	1, 1	
Required distance $= 2\ln 2 - 1$	1	The question asks for the distance from the plane $x = 1$ and the answer is required in cm.
$\approx \underline{39 \text{ cm}}$	1	

8 CIRCULAR MOTION

Answer	Mark	Examiner's tip
1 (i) 33 rev min^{-1} = 66π rad min^{-1}		To convert revs to rads, multiply by 2π.
$= 66\pi \div 60 = \underline{3.46 \text{ rad s}^{-1}}$	1	To convert min^{-1} to s^{-1}, divide by 60.
(ii) $v = r\omega$ $\therefore v = 34.6 \text{ cm s}^{-1} = \underline{0.346 \text{ m s}^{-1}}$	1, 1	Use $v = r\omega$ and change the units.
(iii) $\dfrac{v^2}{r} = 0.346^2 \div 0.1 = 1.197 \text{ m s}^{-2}$	1, 1	Use $\dfrac{v^2}{r}$ with care over units.
radially inwards.	1	The central acceleration produces the circular motion.
2 		No acceleration in the vertical direction.
($\uparrow$) $R \cos \alpha = mg$	1	
($\rightarrow$) $R \sin \alpha = \dfrac{m \times 20^2}{120}$	1	Force $= \dfrac{mv^2}{r}$, radially.
$\therefore$ $\tan \alpha = \dfrac{20^2}{120g} = \dfrac{1}{3}$	1	Divide one equation by the other and use $g = 10 \text{ m s}^{-2}$.
$\therefore$ $\underline{\alpha = 18.4°}$	1	Find α.

Answer	Mark	Examiner's tip

3

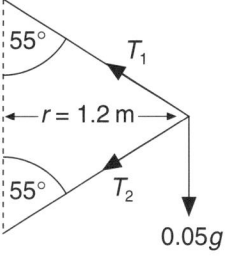

$$T_1 \cos 55° = 0.05g + T_2 \cos 55°$$

2   Resolve vertically with $g = 9.81\,\mathrm{m\,s^{-2}}$.

$$T_1 \sin 55° + T_2 \sin 55° = \frac{0.05 \times 25}{1.2}$$

2   Resolve horizontally and use Force = mass × acceleration.

∴ $T_1 - T_2 = 0.855$

Simplify equations before solving.

$T_1 + T_2 = 1.272$ 1

$\underline{T_1 = 1.06\,\mathrm{N},\ T_2 = 0.21\,\mathrm{N}}$ 1, 1   Obtain T_1 and T_2.

4 (i)

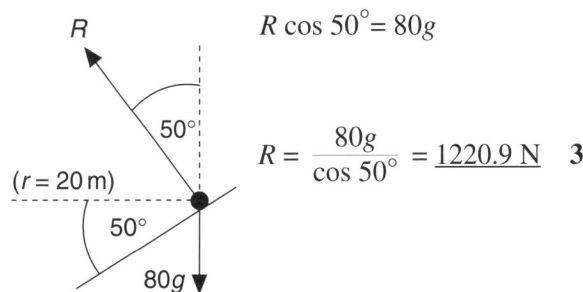

$R \cos 50° = 80g$

Resolve vertically as there is no acceleration component, and use $g = 9.81\,\mathrm{m\,s^{-2}}$.

$$R = \frac{80g}{\cos 50°} = \underline{1220.9\,\mathrm{N}} \quad 3$$

(ii)   Mass × acceleration $= R \sin 50°$

∴   acceleration $= 11.7\,\mathrm{m\,s^{-2}}$ 3

Resolve horizontally using Newton's Second Law.

(iii) Acceleration $= \dfrac{v^2}{r}$

Use acceleration $= \dfrac{v^2}{r}$.

∴ $v^2 = 20 \times 11.7$

$\underline{v = 15.3\,\mathrm{m\,s^{-1}}}$ 2

$ma = mr\omega^2 = R \sin 50°$

Taking particular values for the radius or making assumptions about mass will not result in a complete solution.

$mg = R \cos 50°$

$\omega^2 = \dfrac{g}{r} \tan 50°$ 2

Find angular speed which is independent of mass and inversely proportional to the square root of the radius. So a smaller radius implies bigger angular speed and a shorter time.

As r become larger, ω becomes less, so the time is longer.

$\underline{C \text{ takes the shorter time.}}$ 2

5 (i)

Measure potential energy from the lowest point.

Energy $\frac{1}{2}mU^2 = mga$ 2   Use Conservation of Energy.

$U^2 = 2ga$

$\underline{U = \sqrt{2ga}}$ 2

(continued)

Answer	Mark	Examiner's tip

(ii) 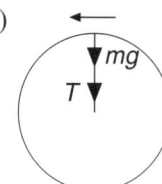 At top, $T \geq 0$

$\therefore mg = \dfrac{mv^2}{a}$ if $T = 0$. **2**

Energy: $\frac{1}{2}mU^2 = \frac{1}{2}mv^2 + 2mga$

$\therefore U^2 = v^2 + 4ga = 5ga \Rightarrow \underline{U = \sqrt{5ga}}$ **2**

The condition for complete circles is that $T \geq 0$ for a string.

Use Newton's Second Law radially.

Use Conservation of Energy.

Eliminate v between the two equations.

(iii) 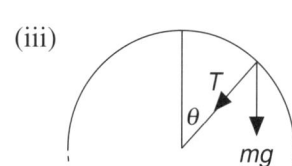 $T + mg \cos \theta = \dfrac{mv^2}{a}$

Particle leaves path when $T = 0$ **2**

$\therefore v^2 = ga \cos \theta$ **2**

Newton's Second Law in radial direction.

This will occur above horizontal position so measure θ from the upward vertical.

Conservation of Energy:

$\frac{1}{2}mv^2 + mg(a + a \cos \theta) = \frac{1}{2}mU^2$

$\therefore mga + \frac{3}{2}mga \cos \theta = \frac{7}{4}mga \Rightarrow \cos \theta = \frac{1}{2}$ **2**

$\therefore$ leaves path at $60°$ to upward vertical. **2**

Conservation of Energy equation.

Solve the equations to find θ.

Explain your answer in words.

(iv) $v^2 = \frac{1}{2}ga$, $v = \sqrt{\frac{1}{2}ga}$ **2**

Substitute θ into the expression for v.

6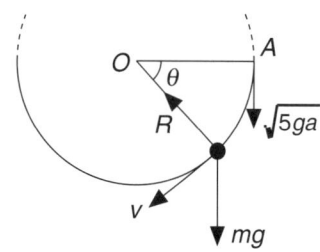

(a) Energy :

$\frac{1}{2}m5ga = \frac{1}{2}mv^2 - mga \sin \theta$ **3**

$\therefore \quad \underline{v^2 = ag(5 + 2 \sin \theta)}$ **1**

Use Conservation of Energy to equate initial energy with energy after subsequent motion.

(b) $R - mg \sin \theta = \dfrac{mv^2}{a}$ **2**

$R = mg \sin \theta + mg(5 + 2 \sin \theta)$ **1**
$\underline{R = mg (5 + 3 \sin \theta)}$ **1**

Use Newton's Second Law radially.

(c) $R \geq 2mg$ for all θ **2**

As $R \neq 0$ at any point the marble remains in contact.

Consider whether $R = 0$ and as $\sin \theta$ has minimum value -1, R has minimum value $2mg$ so cannot be zero.

(d) Put $\sin \theta = -1$ **1**

$\underline{v_{min} = \sqrt{3ag}}$ **1**

(e) Friction and air resistance. **2**

9 SIMPLE HARMONIC MOTION

Answer	Mark	Examiner's tip

1 (a) Amplitude $= 0.025\,\text{m}$

Mark: **1** — Distance from one end of oscillation to the other is $2 \times$ amplitude.

$$v_{max} = a\omega \Rightarrow 20 = 0.025\omega$$

Mark: **1** — Maximum velocity is when $x = 0$ and $v_{max} = a\omega$.

$$\therefore \quad \omega = 800$$

Mark: **1**

Number of oscillations $= \dfrac{800}{2\pi} = 127$

Mark: **1, 1** — The period of an S.H.M. is $\dfrac{2\pi}{\omega}$ and the number of oscillations per second is 1/period.

(b) $F_{max} = ma\omega^2$

Mark: **1** — The acceleration in S.H.M. is $-\omega^2 x$.

$$= 0.2 \times 0.025 \times 800^2$$

Mark: **1**

$$= 3200\,\text{N}$$

Mark: **1** — The maximum acceleration is when $x = a$ and the force is thus $ma\omega^2$.

2 (i) $x = b \cos \omega t$

Mark: **1, 1** — Use $\cos \omega t$ as motion starts from an end point. Use b as amplitude since $x = b$ when $t = 0$.

(ii) $v = \sqrt{\omega^2(b^2 - x^2)} = \sqrt{\omega^2(b^2 - a^2)}$

Mark: **1** — 'Write down' implies that this result can be used without derivation.

$$\therefore \quad b = \sqrt{a^2 + \dfrac{v^2}{\omega^2}}$$

Mark: **1**

(iii) *Either:*

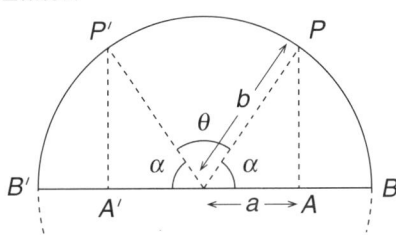

The first method shown here uses the auxiliary circle. As the particle moves from A to A' consider another particle moving from P to P' with angular speed ω.

$$\theta = \pi - 2\alpha, \; \alpha = \cos^{-1}\left(\dfrac{a}{b}\right)$$

Mark: **1**

$$\therefore \theta = \pi - 2\cos^{-1}\left(\dfrac{a}{b}\right)$$

Mark: **1**

$\therefore$ Time from A to A' = time from P to P'

$$= \dfrac{\theta}{\omega} = \dfrac{1}{\omega}\left(\pi - 2\cos^{-1}\dfrac{a}{b}\right)$$

Mark: **1** — ($\cos^{-1}$ = arccos)

(continued)

Answer	Mark	Examiner's tip

Or: use $a = b\cos\omega t_1$
$$-a = b\cos\omega t_2$$

Time $= t_2 - t_1$ **1**   This second method uses the equation from (i) and solves the trigonometric equations to obtain an answer.

$$= \frac{1}{\omega}\cos^{-1}\left(-\frac{a}{b}\right) - \frac{1}{\omega}\cos^{-1}\left(\frac{a}{b}\right)$$ **1**

$$= \frac{1}{\omega}\left(\pi - 2\cos^{-1}\left(\frac{a}{b}\right)\right)$$ **1**

(iv) Second particle takes $\dfrac{\pi}{\omega}$, so   Note the time taken to travel from A to A' is half the period of the second motion.

difference is $\dfrac{2}{\omega}\cos^{-1}\left(\dfrac{a}{b}\right)$ **1**

(v) $\dfrac{\pi}{\omega} = 0.1$ so $\omega = 10\pi$ **1**   Uses half period (rest to rest) = 0.1.

Using (ii) gives $b = \sqrt{0.2^2 + \dfrac{4}{100\pi^2}}$

$$\approx 0.2098$$ **1**

Using (iv) gives $\dfrac{2}{10\pi}\cos^{-1}\left(\dfrac{0.2}{0.2098}\right)$ **1**   Ensure that you work in radians.

$$= 0.0195\ldots = \underline{0.02\,\text{s}}$$ **1**

3 (a) (i)   This is a standard simple pendulum question.

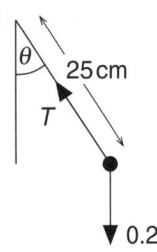

Using Newton's Law perpendicular to the string: **1**   Use Newton's Second Law perpendicular to the string with acceleration $= r\ddot{\theta}$.

$-0.2g\sin\theta = 0.2 \times 0.25\,\ddot{\theta}$ **1, 1**   Take $g = 10\,\text{m}\,\text{s}^{-2}$.

(ii) As θ is small $\sin\theta \approx \theta$ S.H.M. equation with $\omega^2 = 40$.

$\therefore \ddot{\theta} = -40\theta \Rightarrow \underline{\omega^2 = 40}$ **1**

(b) Period $= \dfrac{2\pi}{\omega}$ used   **1**   Use formula which must be learned.

$$= \underline{0.993\ \text{s}}$$ **1**

10 ELASTIC STRINGS AND SPRINGS

Answer	Mark	Examiner's tip

1 Initial energy $= \dfrac{1}{2}\dfrac{\lambda x^2}{l} = \dfrac{39 \times 0.7^2}{1.5}$ **1**

As motion is horizontal the potential energy is unchanged, so need not be considered.

Final energy $= \dfrac{1}{2} \times 0.13v^2$ **1**

Using Conservation of Energy.   **1**
$v = 14\,\text{m s}^{-1}$

The initial elastic energy is equated to the final kinetic energy, and v is made the subject of the formula.

2 'light'   **1**

$$0.2g = \frac{\lambda e}{l} = \frac{10e}{0.4}$$ **2**

Use Hooke's Law for tension and resolve vertically using $g = 9.8\,\text{m s}^{-2}$.

$$\therefore \quad e = \frac{0.2 \times 9.8 \times 0.4}{10} = 0.0784$$

$\therefore$ length $OP = 0.4784$ **1**

Calculate $e + l$.

3 (a) $l = 1.5$, $\lambda = 8 \times 1.5$, or $k = 8$

Modulus = stiffness $\times$ natural length. This examining board uses stiffness, instead of modulus, and $g = 10\,\text{m s}^{-2}$.

At O:
K.E. $= 0$, P.E. $= 0.5g(x + 1.5)$, E.E. $= 0$
At A:

K.E. $= 0$, P.E. $= 0$, E.E. $= \dfrac{1}{2} \times \dfrac{8 \times 1.5x^2}{1.5}$ **2**

Learn the formula for elastic potential energy(E.E.):

E.E. $= \dfrac{1}{2} \times \dfrac{\lambda x^2}{l}$ or $\dfrac{1}{2}kx^2$

Conservation of Energy gives
$$5(x + 1.5) = 4x^2$$
$$\therefore \; 8x^2 - 10x - 15 = 0$$ **2**

Conservation of Energy simplifies the working in many questions.

(b) $x = \dfrac{10 \pm \sqrt{100 + 480}}{16} = 2.13$

Solve quadratic using the quadratic formula and select the value for which $x > 0$.

$\therefore OA = l + x = 3.63$ m (to 3 s.f.)   **3**

OA = natural length + extension.

4 (i) Elastic potential energy $= \dfrac{1}{2} \times \dfrac{\lambda x^2}{l}$

Use formula for elastic potential energy.

$$= \dfrac{1}{2} \times \dfrac{32x^2}{0.8} = 20x^2$$ **2**

(ii) Use Conservation of Energy.
At top,
K.E. $= 0$, E.E. $= 0$, P.E. $= 0.5g(0.8 + x)$

Find each energy term at start.

At lowest point,
K.E. $= 0$, E.E. $= 20x^2$, P.E. $= 0$

Find each energy term at lowest point.

$\therefore 20x^2 = 4.9x + 3.92$ **2**

Equate energies.

$20x^2 - 4.9x - 3.92 = 0$, $\therefore \; x = 0.58\,\text{m}$ **2**

Rearrange to give a quadratic equation and solve to find x.

(*continued*)

Answer	Mark	Examiner's tip

(iii) When $x = 0.1$:

K.E. $= \dfrac{1}{2} \times 0.5v^2$, E.E. $= 0.2$, P.E. $= 0$. **2**

At start:

K.E. $= 0$, E.E. $= 0$, P.E. $= 0.5g \times 0.9$ **1**

$\therefore \quad 0.45g = 0.25v^2 + 0.2$

$\therefore \qquad v^2 = 16.84 \Rightarrow \underline{v = 4.1\,\text{m s}^{-1}}$ **2**

5 (a) $150 = \dfrac{\lambda \times 0.1}{0.2}$ **2**

$\underline{\lambda = 300\,\text{N}}$ **1**

(b) $AP^2 = 15^2 + 20^2$ **1**

$\therefore \; AP = 25\,\text{cm}$ **1**

$\therefore \quad T = \dfrac{300 \times 0.3}{0.2} = 450\,\text{N}$ **2**

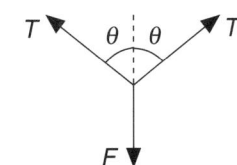

$\therefore \; F = 2T \cos\theta$ **1**

$\quad = 2 \times 450 \times \dfrac{4}{5}$

$\quad = \underline{720\,\text{N}}$ **1**

(c) Original energy is elastic $= \dfrac{1}{2}\dfrac{\lambda \times (0.3)^2}{0.2}$ **2**

$\therefore \; \dfrac{1}{2} \times 0.1 \times v^2 = \dfrac{300}{2 \times 0.2}(0.3^2 - 0.1^2)$ **2**

$\underline{\text{Speed} = 34.6\,\text{m s}^{-1}}$ **2**

6 (a)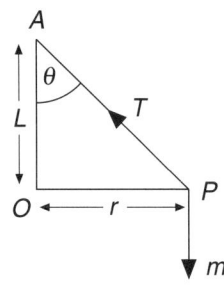

($\uparrow$) $T \cos\theta = mg$ **2**

$T = \dfrac{4mgx}{L}$ **1**

$\cos\theta = \dfrac{L}{L+x}$ **2**

Examiner's tip column:

Here $g = 9.8\,\text{m s}^{-2}$.

Take zero potential energy $0.9\,\text{m}$ below O this time.

Find each energy term at start and at required point.

Equate energies and rearrange to find v.

Hooke's Law.

First find new extension so that new tension can be calculated.

String is now $50\,\text{cm}$ long and extension is $30\,\text{cm} = 0.3\,\text{m}$.

Then resolve perpendicular to AB.

Use Conservation of Energy. As the motion is in a horizontal plane there is no gravitational potential energy term, so gain in kinetic energy equals loss of elastic energy.

There are 3 unknowns; T, θ and x, the extension, so 3 equations are needed.

Resolve vertically, use Hooke's Law and use trigonometry with length $AP = L + x$.

Answer	Mark	Examiner's tip

Solving $\cos\theta = \dfrac{L}{4x} = \dfrac{L}{L+x}$ **1**   Eliminate T and θ to find x.

$\therefore \quad x = \dfrac{L}{3}$ **1**

(b) $r = \sqrt{\left(\dfrac{4L}{3}\right)^2 - L^2} = \dfrac{L}{3}\sqrt{7}$ **2**   Use Pythagoras' theorem to find the radius of the circular motion.

$T = \dfrac{4mg}{3}$ **1**   Find the value of T by using $x = \dfrac{L}{3}$.

$T\sin\theta = \dfrac{mv^2}{r}$ **2**   Resolve radially for the circular motion.

Use $\sin\theta = \dfrac{r}{L+x}$ and substitute   **1**   Show all your working clearly as the answer is printed.

$\Rightarrow \quad v = \dfrac{1}{3}\sqrt{7gL}$ **1**

7 (a) W.D. $= \displaystyle\int_0^l \dfrac{\lambda x}{l}\,dx = \left[\dfrac{\lambda x^2}{2l}\right]_0^l$ **1**   Work done (W.D.) $= \displaystyle\int_{x_1}^{x_2} F\,dx$

$= \dfrac{\lambda l}{2}$ **1**   – the force here is tension, x increases from 0 to l.

At the start: K.E. $= 0$, E.E. $= 0$, P.E. $= mg.2a$ Use Conservation of Energy. Take zero potential energy at point being considered – here the lowest point.

At the lowest point: K.E. $= 0$, P.E. $= 0$

$\therefore mg2l = \dfrac{\lambda l}{2} \Rightarrow \underline{\lambda = 4mg}$ **1, 1**

(b) Maximum speed $\Rightarrow$ acceleration $= 0$ Maximum speed means $\dfrac{dv}{dt} = 0$, i.e. acceleration $= 0$.

$\therefore T = mg \Rightarrow \dfrac{4mgx}{l} = mg \Rightarrow x = \dfrac{l}{4}$ **1, 1**   Use Hooke's Law for tension and equate to mg.

At this point K.E. $= \dfrac{1}{2}mv^2$, P.E. $= 0$,

$\text{E.E.} = \dfrac{1}{2}\times\dfrac{4mg}{l}\left(\dfrac{l}{4}\right)^2.$

At start, P.E. $= mg\dfrac{5l}{4}$, K.E. $= 0$, E.E. $= 0$.   Again use energy but the zero P.E. is now chosen at the new point under consideration.

$\therefore \dfrac{1}{2}mv^2 + \dfrac{1}{2}\times\dfrac{4mg}{l}\left(\dfrac{l}{4}\right)^2 = \dfrac{5mgl}{4}$ **1, 1**

$\therefore \quad v = \dfrac{3}{2}\sqrt{gl}$ **1**

(*continued*)

Answer	Mark	Examiner's tip
(c) Distance $= l + \dfrac{l}{4} = \dfrac{5l}{4}$	1	A is the equilibrium position which was found earlier.
(d) (i) $m\ddot{y} = mg - \dfrac{\lambda\left(\dfrac{l}{4} + y\right)}{l}$	1	This method should be learned as it is a standard piece of bookwork.
$\qquad = \dfrac{-\lambda y}{l} = \dfrac{-4mgy}{l}$	1	
$\qquad \therefore \ \ddot{y} = \dfrac{-4gy}{l}$	1	
(ii) $\omega^2 = \dfrac{4g}{l}$		
$\qquad \therefore \text{Period} = 2\pi\sqrt{\dfrac{l}{4g}} = \pi\sqrt{\dfrac{l}{g}}$	1	Formula for period, $\dfrac{2\pi}{\omega}$, should be learned.
(e) Natural length of rope.	1	The period is independent of amplitude.
Period remains constant.	1